HOLISTIC APPROACHES TO INFECTIOUS DISEASES

HOLISTIC APPROACHES TO INFECTIOUS DISEASES

Edited by

Anne George, MD

Joshy K. S.

Mathew Sebastian, MD

Oluwatobi Samuel Oluwafemi, PhD

Sabu Thomas, PhD

AAP | APPLE ACADEMIC PRESS

Apple Academic Press Inc.
3333 Mistwell Crescent
Oakville, ON L6L 0A2
Canada

Apple Academic Press Inc.
9 Spinnaker Way
Waretown, NJ 08758
USA

Exclusive worldwide distribution by CRC Press, a member of Taylor & Francis Group

Printed in the United States of America on acid-free paper
International Standard Book Number-13: 978-1-77188-312-2 (Hardcover)
International Standard Book Number-13: 978-1-77188-313-9 (eBook)

Library and Archives Canada Cataloguing in Publication

Holistic approaches to infectious diseases / edited by Anne George, MD, Joshy K. S. Mathew Sebastian, MD, Oluwatobi Samuel Oluwafemi, PhD, Sabu Thomas, PhD.

Includes bibliographical references and index.
Issued in print and electronic formats.
ISBN 978-1-77188-312-2 (hardcover).--ISBN 978-1-77188-313-9 (pdf)

1. HIV infections. 2. AIDS (Disease). 3. Communicable diseases. 4. Medicine, Ayurvedic. I. Thomas, Sabu, editor II. Sebastian, Mathew, editor III. George, Anne, 1961-, editor IV. Oluwafemi, Oluwatobi Samuel, editor V. K. S., Joshy, editor

RA643.8.H64 2016 362.19697'92 C2016-903811-4 C2016-903812-2

Library of Congress Cataloging-in-Publication Data

Names: George, Anne, 1961- editor. | Joshy K. S., editor. | Sebastian, Mathew, editor. | Oluwafemi, Oluwatobi Samuel, editor. | Thomas, Sabu, editor.
Title: Holistic approaches to infectious diseases / editors, Anne George, Joshy K.S., Mathew Sebastian, Oluwatobi Samuel Oluwafemi, Sabu Thomas.
Description: Toronto ; New Jersey : Apple Academic Press, 2016. | Includes bibliographical references and index.
Identifiers: LCCN 2016025236 (print) | LCCN 2016025775 (ebook) | ISBN 9781771883122 (hardcover : alk. paper) | ISBN 9781771883139 (eBook) | ISBN 9781771883139 ()
Subjects: | MESH: HIV Infections--therapy | Medicine, Ayurvedic
Classification: LCC RC111 (print) | LCC RC111 (ebook) | NLM WC 503.2 | DDC 616.9--dc23
LC record available at https://lccn.loc.gov/2016025236

Apple Academic Press also publishes its books in a variety of electronic formats. Some content that appears in print may not be available in electronic format. For information about Apple Academic Press products, visit our website at **www.appleacademicpress.com** and the CRC Press website at **www.crcpress.com**

ABOUT THE EDITORS

Anne George, MD

Associate Professor, Department of Anatomy, Government Medical College, Kottayam, India

Anne George, MD, is an Associate Professor at the Government Medical College, Kottayam, Kerala, India. She did her MBBS Bachelor of Medicine and her Bachelor of Surgery at Trivandrum Medical College, University of Kerala, India. She acquired a DGO (Diploma in Obstetrics and Gynaecology) from the University of Vienna, Austria; a Diploma of Acupuncture from the University of Vienna; and her MD from Kottayam Medical College, Mahatma Gandhi University, Kerala, India. She has organized several international conferences, is a fellow of the American Medical Society, and is a member of many international organizations. She has five publications to her name and has presented 25 papers.

Joshy K. S.

Joshy K. S., is a researcher working at the International and Inter University Centre for Nanoscience and Nanotechnology and C.M.S. College, Mahatma Gandhi University, Kottayam, Kerala, India. She has received a bachelor's degree in chemistry from Kerala University, Kerala, India; a master's degree in polymer chemistry from the School of Chemical Sciences Mahatma Gandhi University, Kottayam, Kerala, India; and an MPhil degree in polymer chemistry from Cochin University of Science and Technology, Kochi, Kerala, India. Joshy carried out her PhD work at Mahatma Gandhi University, Kottayam, Kerala, India. She is currently pursuing her research on the development of lipid and polymer nanoparticles for drug delivery applications. She has published in international journals and conference proceedings.

Mathew Sebastian, MD

Senior Consultant Surgeon, Elisabethinin Hospital, Klagenfurt, Austria; Austrian Association for Ayurveda

Mathew Sebastian, MD, has a degree in surgery (1976) with a specialization in Ayurveda. He holds several diplomas in acupuncture, neural therapy (pain therapy), manual therapy, and vascular diseases. He was a missionary doctor in Mugana Hospital, Bukoba in Tansania, Africa (1976–1978) and underwent surgical training in different hospitals in Austria, Germany, and India for more than 10 years. Since 2000 he has been the doctor in charge of the Ayurveda and Vein Clinic in Klagenfurt, Austria. At present he is a Consultant Surgeon at Privatclinic Maria Hilf, Klagenfurt. He is a member of the scientific advisory committee of the European Academy for Ayurveda, Birstein, Germany, and TAM advisory committee (Traditional Asian Medicine, Sector Ayurveda) of the Austrian Ministry for Health, Vienna. He conducted an International Ayurveda Congress in Klagenfurt, Austria, in 2010. He has several publications to his name.

Oluwatobi Samuel Oluwafemi, PhD

Senior Lecturer, Department of Chemistry and Chemical Technology, Walter Sisulu University, Mthatha Campus, Eastern Cape, South Africa

Oluwatobi Samuel Oluwafemi, PhD, is a Professor at the Department of Applied Chemistry, University of Johannesburg, South Africa. He has published many papers in internationally reviewed journals and has presented at several professional meetings. He is a fellow of many professional bodies and is a reviewer for many international journals, and he has received many awards for his excellent work in material research. His current research interests include green synthesis and application of nanoparticles in medicine, water treatment, polymer, LEDs, and sensors.

Sabu Thomas, PhD

Director, International and Inter University Centre for Nanoscience and Nanotechnology, Mahatma Gandhi University, Kottayam, Kerala, India

Dr. Sabu Thomas is a Professor at the School of Chemical Sciences and Honorary Director of the International and Inter University Centre for Nanoscience and Nanotechnology, Mahatma Gandhi University, Kottayam, Kerala, India. He joined Mahatma Gandhi University as a full-time faculty in 1987. He has been associated with several universities in Europe, China, Malaysia, and South Africa. Professor Thomas is a member of the Royal Society of Chemistry of London, a member of the New York Academy of Science, USA, and the recipient of awards from the Chemical Research Society of India and the Material Research Society of India (2013). Professor Thomas has supervised 65 PhD theses, and he has more than 530 publications, 43 books, four patents, and 18163 citations to his credit. The h-index of Prof. Thomas is 68, and he is listed as the 5th position in the list of Most Productive Researchers in India in 2008. His research focuses on polymer blends, recyclability, reuse of waste plastics and rubbers, fiber-filled polymer blends, nanocomposites, elastomers, pervaporation phenomena, and sorption and diffusion.

CONTENTS

LIST OF CONTRIBUTORS

Sana Aboobacker
Department of Pharmacology, Fr. Muller Medical College, Mangalore, Karnataka, India – 575002

Beena Antony
Fr. Muller Medical College, Mangalore, Karnataka, India – 575002

Mona Bakr
The National Institute for Laser Enhanced Sciences, Cairo University, Egypt

Samarth Bhatt
Jena University Hospital, Friedrich Schiller University, Institute of Human Genetics, Kollegiengasse 10, D-07743 Jena, Germany

Purna Chandra Dash
Health Economist, Team Leader, Conseil Sante, France

Tarek A. El-Tayeb
The National Institute for Laser Enhanced Sciences, Cairo University, Egypt

Mini Ghosh
School of Advanced Sciences, VIT University, Chennai Campus, India

Iman E. Gomaa
German University in Cairo, Egypt

Poonam Gupta
Molecular Virology Laboratory, Department of Biotechnology, Jamia Millia Islamia, New Delhi – 110025, India, E-mail: superm12@gmail.com

Mohammad Husain
Molecular Virology Laboratory, Department of Biotechnology, Jamia Millia Islamia, New Delhi – 110025, India, E-mail: mhusain2@jmi.ac.in

Mwiya Liamunga Imasiku
Department of Psychiatry, School of Medicine, University of Zambia, Lusaka, Zambia

Thomas Liehr
Jena University Hospital, Friedrich Schiller University, Institute of Human Genetics, Germany

R. Oviya
Department of Bioinformatics, Bharathiar University, Coimbatore – 641046, India

Princy Louis Palatty
Fr. Muller Medical College, Mangalore, Karnataka – 575002, India

P. Rajendran
Programme Director, PACT Programme, HLFPPT, Bhopal, India

R. Sathishkumar
Department of Biotechnology, Salem Sowdeswari College, Salem – 636010, India

M. Sharanya
Department of Bioinformatics, Bharathiar University, Coimbatore – 641046, India

Surya Pratap Singh
Department of Biochemistry, Banaras Hindu University, Varanasi, India

Srinadh
M&E Officer, PACT Programme, HLFPPT, Bhopal, India

Daniela Cristina Stefan
Department of Pediatrics and Child Health, Tygerberg Children's Hospital Stellenbosch University, Cape Town, South Africa

Susan Westfall
Department of Biochemistry, Banaras Hindu University, Varanasi, India; Department of Pharmaceutics, Faculty of Pharmacy, Babu Banarasi Das National Institute of Technology & Management, Lucknow – 227105, Uttar Pradesh, India

LIST OF ABBREVIATIONS

Aβ	amyloid beta
AD	Alzheimer's disease
ADL	actions of daily living
AIDS	acquired immune deficiency syndrome
ALL	acute lymphoblastic leukemia
APP	amyloid precursor protein
ART	antiretroviral treatment
ARVs	anti-retroviral drugs
BL	Burkitt lymphoma
BOR	bed occupancy rate
BSS	behavioral surveillance survey
BTR	bed turnover ratio
CAE	*Centella asiatica* extract
CBAs	carbohydrate-binding agents
CCC	community care centre
CCE	continued comprehensive education
CCR	chemokine coreceptor
CIN	cervical intraepithelial neoplasia
CMIS	computerized management information system
CMV	cytomegalovirus
CRFs	circulating recombinant forms
CRI	co-receptor inhibitor
CSF	colony-stimulating factors
CSSP	center for specialized studies and programs
DAR	drug adherence rate
DTC	diethyldithiocarbamate
EBV	Epstein Barr virus
ETC	electron transport chain
EWBS	existential well-being scale
FDA	Food and Drug Administration
FIs	fusion inhibitor
GNA	*galanthus nivalis* agglutinin
HAART	highly active antiretroviral therapy

HD	Huntington's disease
HHA	*hippeastrum* hybrid agglutinin
HIV	human immunodeficiency virus
HLFPPT	Hindustan Late Family Planning Promotion Trust
HPV	human papilloma virus
HSCs	hematopoietic stem cells
HSP	heat shock protein
HSV	herpes simplex virus
IDUs	intravenous drug users
IL-6	interleukin-6
INSTI	integrase strand transfer inhibitor
IRIS	immune reconstitution inflammatory syndrome
KS	Kaposi sarcoma
LED	light-emitting diode
mAb	monoclonal antibody
MDGs	millennium development goals
MFC	minimal fungicidal concentration
MFD	minimum fungicidal dilution
MIC	minimal inhibitory concentration
MID	minimum inhibitory dilution
MIS	management information system
MS	mass spectroscopy
MSM	men having sex with men
MVA	modified vaccinia virus Ankara
NACP	National Aids Control Program
NCI	National Cancer Institute
NFHS	National Family Health Survey
NHL	non-Hodgkins lymphoma
NIH	National Institutes of Health
NILES	National Institute for Laser Enhanced Sciences
NMR	nuclear magnetic resonance
NNRTIs	non-nucleoside reverse transcriptase inhibitors
NO	nitric oxide
NPs	nanoparticles
NtRTIs	nucleotide reverse transcriptase inhibitors
OH^-	hydroxyl radical
OIs	opportunistic infections
PACT	promoting access to care and treatment

PAF	platelet-activating factor
PBS	phosphate buffered saline
PCNSL	primary CNS lymphomas
PCP	*Pneumocystis jirovecii* pneumonia
PD	Parkinson's disease
PIs	protease inhibitors
PRM-A	Pradimic In a
QPRs	quarterly progress reports
RISC	RNA-induced silencing complex
ROS	reactive oxygen species
RWBS	religious wellbeing scale
SCC	squamous cell carcinoma
SDA	Sabourauds dextrose agar
SDAT	senile dementia of Alzheimer's type
siRNA	small interfering RNA
SIVs	simian immunodeficiency viruses
SR	sub-recipient
SWBS	spiritual well-being scale
TB	tubercle bacillus
TCM	traditional chinese medicine
TIs	targeted interventions
TMPMP	trimethoxy phenyl 1' methoxy propionaldehyde
TNF	tumor necrosis factor
UDA	*Urtica dioica* agglutinin
WHO	World Health Organization

PREFACE

This book describes a holistic approach to the prevention and control of infectious diseases from enteric pathogens. Holistic approaches to infectious diseases deal with different concepts or approaches to take care of the challenging diseases. According to the World Health Organization reports, infectious and parasitic diseases are the second leading cause of death in the world, and the leading cause of infectious disease is due to the enteric pathogens and they cause almost two million deaths every year. The first four chapters of the book deal with different approaches such as ayurvedic, bioinformatic, fungal, and metal-based treatment to diseases. The remaining chapters fully focus on various approaches to HIV and AIDS—one of the most challenging infectious disease to mankind.

Ancient medical practices provide novel paradigms to understand chronic neurodegenerative diseases due to their unique outlook of disease pathology. Ayurveda is the ancient medical wisdom of India that provides novel and powerful insights for the preemptive and therapeutic treatment of neurodegenerative diseases. Ayurveda describes the disease pathology as being derived from the accumulation of *vata* or air in the brain. There are also traditional formulations composed of several herbs whose combined action against various aspects of the disease pathology effectively rescues the symptoms of the disease. It has been shown that natural therapies are equally or more effective than modern medicines with a longer efficacy and zero side effects.

Various microscopic organisms live harmlessly inside the body and on the surface of the skin. However, certain types of fungus that are normally harmless, on overgrowth, can cause superficial and systemic infections, which are more commonly seen in those people undertaking antibiotics, corticosteroids, immunosuppressant drugs and contraceptives. This also prevails in people with endocrine disorders, immune diseases, diabetes and others diseases, such as AIDS, tuberculosis, major burns and leukemia, especially, found in obese people with excessive skin folds.

Alternative medicine is an age-old tradition and a proven skill which is presently entering the mainstream, holding promises for combating various dreadful diseases in which our modern medicines have failed.

Though traditional medicine has been developed more by observation, practice and skill rather than scientific proven mechanism, still, the medicinal effectiveness is too broad and effective to be ignored. Recent advances in holistic medicinal research are towards uncovering the molecular mechanism of inhibition. Advances in biology and biotechnology have greatly improved our understanding of the human system and mode of drug-target interaction. To aid with these technologies, computational approaches termed bioinformatics play an important role. Genomics, proteomics, transcriptomics, data mining, text mining, network construction, expression studies, etc. are some areas of bioinformatics which when interconnected interpret various important information on the medical aspects and opens up challenges and solutions towards personalized medicine and drug discovery.

From Chapter 5 onwards the chapters concentrate on HIV and AIDS and its recent trends and treatment regimes, case studies, etc. The Acquired Immune Deficiency Syndrome (AIDS) is a deadly disease of the human immune system that is caused by infection with the human immunodeficiency virus (HIV). As per the current understanding of AIDS, in the initial stage of infection, a person experiences a brief period of influenza-like illness. Later in the subsequent stages, this is typically followed by a prolonged period without any significant symptoms. As the illness progresses, it interferes more and more with the immune system, making the person much more likely to get infections, including opportunistic infections OIs and tumors that do not usually affect people who have efficient working immune systems. Worldwide, about 33.2 million people live with AIDS and about 2.1 million AIDS-related deaths occur each year including 3,30,000 children. Though a breakthrough has been reported, so far there is no cure for AIDS, and this disease is endemic in many parts of the world and especially in sub-Saharan Africa. The old wisdom of 'prevention is better than cure' is truly applicable in the case of HIV/AIDS transmission. The most common HIV defining and related cancers in children are: Kaposi sarcoma (KS), Burkitt lymphoma (BL) and leyomyosarcoma (mainly in developed countries). Despite progress in the treatment and survival of childhood cancer globally, the need for adapted protocols and randomized trials for HIV-related malignancies in children remains a priority. The awareness and impact of HIV/AIDS is different in developed countries and developing countries. Various studies show that about 75% of infected people are aware of their HIV sero status in the USA and Europe, while only 10–20% in Indians. There are many factors attributing to this, which include limited access to health care facilities by the population, lack of basic infrastructure for early diagnosis of HIV, which requires well-established laboratories, lack of funds to support awareness or preventive campaigns as well as the specific treatment like highly active antiretroviral therapy (HAART). As reported by certain investigators, the spectrum of OI and the prevailing HIV sero-status are expected to change as a result of HAART. Hence an early detection and diagnosis of OI may help in effective disease management. Antiretroviral therapy mainly consists of drugs that are capable of reducing the disease burden on the infected individuals and helps in preventing opportunistic infections that are often the cause of death among HIV patients. The groups of drugs range from entry inhibitors to maturation inhibitors, which aim at terminating the further development of HIV by acting at different stages of its life cycle.

The Government of India established a National AIDS Control Program (NACP) to combat the HIV epidemic. During NACP I (1992), there was increased awareness about HIV/AIDS, particularly among the urban population, and subsequent successful intervention programs and the strengthening of STDs clinics across the country were major achievements. During NACP II, the classification of states has focused on the vulnerability of states, with states being classified as high and moderate prevalence and high and moderate vulnerability. The primary goal of NACP III is to halt and reverse the epidemic in India over the next five years by integrating program for prevention, care, support and treatment. Community Care Centre (CCC) is a comprehensive facility-providing medical, counseling, referral and outreach service to the registered PLHIV. The overall goal of the program is to improve the survival and quality of life of people living with HIV/AIDS (PLHIV). Envisioned as a home away from home, CCC is a facility for providing accessible, affordable, and sustainable counseling, support, and treatment to PLHIV. With the medical services being an integral and important part of the program, CCCs have a critical role in helping PLHIV gain easy access to ART treatment and counseling on primary prevention, nutrition, drug adherence, etc. The main functions of these CCCs are to provide treatment to the registered PLHIV on various types of opportunistic infections/side effects, providing quality of counseling, providing home a based care through outreach component and LFU tracking as per the list given by ART Centers. Currently, 35 CCCs are in functional among five states and one union territory of India. A total of 66,471 PLHIV were registered with the CCCs and availing various services till December 2012 for smooth implementation of the program. HLFPPT has been monitoring and providing constant technical support to the Community Care Centers (CCCs) scattered among the above-said states.

In this book, topics on new approaches to infectious diseases, recent trends in HIV AIDS, ongoing treatments, case studies and the major achievements of the Government against this deadly disease are discussed in a lucid manner, and these aspects are presented in a readily accessible form. Each chapter provides an in-depth array of knowledge satisfying persons related to this field like researchers, doctors, students and professionals.

—Sabu Thomas
Anne George
Mathew Sebastian
Oluwatobi Samuel Oluwafemi
Joshy K. S.

CHAPTER 1

THE AYURVEDIC TREATMENT OF NEURODEGENERATIVE DISEASES

SUSAN WESTFALL and SURYA PRATAP

CONTENTS

ABSTRACT

Ancient medical practices provide novel paradigms to understand chronic neurodegenerative diseases due to their unique outlook of disease pathology. Ayurveda is the ancient medical wisdom of India that gives novel and powerful insights for the preemptive and therapeutic treatment of neurodegenerative diseases. Ayurveda describes the disease pathology as being derived from the accumulation of *vata* or air in the brain. In general, treatments include both internal and external practices and herbal therapies to reduce the amount of air in the body in addition to stimulating the central nervous system (*medhya rasayanas*). There are also traditional formulations composed of several herbs whose combined action against various aspects of the disease pathology effectively rescues the symptoms of the disease. Due to the failure of modern allopathic medicine to effectively treat chronic neurodegenerative diseases, many people are turning to alternative natural therapies. It has been shown that natural therapies are equally or more effective than modern medicines with a longer efficacy and zero side effects. The following summarizes the latest research on a battery of Ayurvedic herbs that were traditionally used in the treatment of Parkinson's, Alzheimer's and Huntington's disease and the potential these herbs have to become modern therapies.

1.1 INTRODUCTION

1.1.1 AYURVEDA

Ayurveda is the ultimate holistic medical wisdom and literally means the 'wisdom of life' (*ayus* = life; *veda* = wisdom). It considers the body, mind and spirit to be a seamless system between which proper homeostasis and harmony is required to instill perfect health and a clear path to enlightenment. Health in Ayurveda means not only physical health but, more importantly, mental clarity. Health in Sanskrit is *svastha* which, when broken down means '*sva'* self and '*stha'* established. This self is purposely ambiguous in that it means both the physical 'self' connected to the ego and the higher, spiritual 'Self' that is connected with God. So 'health' in Ayurveda is defined as a proper establishment of self both physically (proper physical health) and mentally (mastery over indulgence, sensory cravings and separation from God). The ultimately goal of Ayurveda is to achieve optimal physical health in order to have a clear path to God and enlightenment.

> *Ayurveda scrutinizes the subtle process of life, studies its nature, ways and conditions of development and deduces there-from a universal course of conduct for man's guidance in life.*
>
> —Ananthacharya (1939)

The God of Ayurveda, Lord *Dhanvantari*, is an avatar of the great Lord *Vishnu* and is known from the Vedas as a physician of the Gods. According to mythology, Lord *Dhanvantari* emerged from the churning of oceans by the *asuras* (demons) and *devas* (Gods) holding a goblet of *amrita*, the almighty nectar of immortality and spiritual enlightenment. Anyone seeking some grace of health will often pray to Lord *Dhavantari*.

In legend, it is said that Ayurveda originated at the base of the Himalayan mountains around 3000 BC during the period of the Indus Valley civilization. It is said that a congregation of the best medical *rishis* at the time convened there to discuss the pathology and treatment of many diseases under the divine guidance of Indra, the God of rain. Indra taught the medicine of the gods, Ayurveda to a missionary of the *rishi*s who since then spread the wisdom of Ayurveda through the generations by oral traditions. Ayurveda is so old that it is even mentioned in all of the four *Vedas*, the sacred texts of Hinduism. Ayurveda was practiced as a traditional medicine for thousands of years until it was finally written down by the master physicians of the time. Even today, the *Charak Samhita* and *Sushruta samhita* remain the ultimate references for Ayurveda, although there are many ancillary texts describing specific treatments using the wisdom of Ayurveda. Ayurveda, even at that time, was divided into eight different sub-specialties namely, *kayachikitsa* (internal medicine), *baala chikitsa* (pediatrics), *graha chikitsa* (demonology), *urdhvanga chikitsa* (diseases of head and neck), *shalya chikitsa* (surgery), *visha chikitsa* (toxicology), *jara chikitsa* (rejuvenation) and *vrsha chikitsa* (aphrodisiac therapy).

In Ayurveda, everything is said to be a unique composition of the five elements or the *pancha maha bhutas*, namely *prithvi* (earth), *aapas* (water), *vayu* (air), *tejas* (fire) and *aakash* (ether). The human body is born with an innate and unique cohort of these elements, which defines one's physical and mental qualities. Physically, the combination of the five elements is referred to as *doshas*, the biological humors. There are three *doshas: vata* (air and ether), *pitta* (fire and water) and *kapha* (earth and water). *Doshas* have no physical bearing but reflect a person's physical tendencies, mental habits and predisposition to reacting to environmental stimuli. More simply, a person's *doshic* inheritance gives them certain physical and mental qualities (i.e., tall versus short, dry versus oily skin, shy versus rambunctious, focused versus aloof temperament, etc.). It is important to note that the *doshic* predisposition that every person is born with, known as *vikriti*, does not change throughout life. Thus, a *vata-pitta* person will always have a *vata-pitta* predisposition. However through life, exposure to unbalancing stimuli can create symptoms of excessive or depleted *doshas* and this state is called *vikriti.* Ultimately, everyone should strive to be a perfect balance of the three *doshas* thus avoids the weaknesses and embrace the strengths of each respective *dosha.*

Very briefly, a *vata* person will have particular characteristics, namely tall and lean with a dreamy and idealistic personality reflecting the elusiveness and changeability of the light air and ether elements. A *pitta* person will be naturally muscular, have a tendency to be hot with a fiery and determined attitude with a predisposition towards aggression. Finally, a *kapha* person will be more solid, slow and steady like their earth element and have a warm, loving and tenacious personality. (Please note that the categorization into a *doshic* type is very deep and these descriptions only barely touch the surface of what characterizes a *dosha*). Nevertheless, a person will be a combination of a defined proportion of each of these *doshas* and hence will have different propensities towards given attributes. This is an important consideration whenever a herb is prescribed to treat a specific disorder. For example, giving a highly heating herb to a person with a lot of internal fire (*pitta*) will disrupt their physiological homeostasis in the treatment of the *vikriti* or disease, likely rearing more disease and disruptions. Likewise, giving an extremely bitter herb to a predominantly *vata* individual will rear many imbalances in their constitution. In this light, the taste (*rasa*), properties (*virya*), qualities (*guna*) and effect of taste after digestion (*vipaka*) need to all be considered before indulging in any herbal treatment.

1.1.1.1 AYURVEDA IN NEURODEGENERATION

One branch of Ayurveda defines the potency of a class of herbs known as the *rasayan chikitsas,* literally 'rejuvenative medicines'. Plants that are considered *rasayan's* are whole-body tonics that stimulate immunity and metabolism and promote health and longevity by optimization of homeostatic processes (Auddy et al., 2003). Such herbs stimulate happiness, youthfulness, divine connectedness and act as tonics for young, middle-aged and elderly alike. Chemically, the magic of these herbs has been postulated to be due to their high content of antioxidants (Sharma et al., 1992), which battle many of the acquired diseases from daily stress, age and environmental toxins. But, not all *rasayan*'s are created equal. There are specific *rasayan*'s that are more adept for targeting particular areas of the body. Of particular interest are the *medhya rasayan*'s, the rejuvenators of the brain. These *rasayan*'s are particularly adept at boosting memory, preventing cognitive deficits and improving overall brain function. From the ancient literature, the most important *medhya rasayan*'s, or nervine tonics, are (in order) Ashwagandha, Brahmi, Jatamansi, Jyotishmati, Mandukparni, Shankhapushpi and Vacha (Ven Murthy et al., 2010).

One of the original Ayurvedic formulations acting as a nervine tonic was actually designed against Parkinson's disease. This mixture contained a mixture of powdered *Hyoscyamus reticulatus* and *Mucuna pruriens* seeds together with *Withania somnifera* and *Sida cordifolia* roots prepared in cow's milk. In a recent clinical trial, the above mixture was administered to 18 clinically diagnosed Parkinsonian patients for 8 weeks and an extensive analysis on their symptomatic improvements was assessed (Nadashayana et al., 2000). This study indicated a significant improvement in the actions of daily living (ADL), motor tasks and an overall improvement in terms of the universal scale of Parkinsonian symptoms, the UPDRS rating. Tremors were reduced and the onset of bradykinesia eradicated. This study attributed the anti-parkinsonian effects to the 200 mg of L-DOPA (the precursor to dopamine) found in each powdered dose, however more recent developments have indicated that there are many other components that can also contribute to its anti-Parkinsonian effects (Katzenschlager et al., 2004; Manyam et al., 2004).

More interestingly, this study went further into the essence of holistic Ayurvedic medicine and included a group of patients who received both the palliative treatment together with traditional cleansing (*panchakarma*) before the onset of treatment. It was found that prior cleansing further enhanced the motor benefits of the palliative treatments and the patients showed a heightened improvement in their ADL and UPDRS scores. The latter indices were only mildly improved in the groups taking the palliative treatments alone. So together, it is not just the active ingredients of the herbs that contribute to the power of Ayurveda, but rather, the entire Ayurvedic paradigm of treating disease (Nadashayana et al., 2000).

1.1.2 NEURODEGENERATION

Neurodegeneration is probably the most feared condition in aging persons. The toll on one's health and family and the level of suffering is incalculable as this voracious disorder silently infects the mental processing of one's most cherished faculty: the brain. There are no treatments, limited understanding and an overall helplessness among doctors and the society to treat aging persons afflicted with this disorder. The economic burden in the US for treating Alzheimer's disease alone supersedes \$100 billion per year. Thus, neurodegeneration is not only a personal problem, but also a societal problem and challenge for politicians and scientists alike.

Neurodegeneration is the progressive loss of neurons due to structural, molecular and functional damage eventually leading to cell death. The main causes of neurodegeneration are age, polyglutamine accumulation, oxidative stress, protein aggregation (proteinopathies) and mitochondrial damage. Of these, the accumulation of oxidative stress from ageing, mitochondrial damage and environmental toxins are the biggest risk factors for neurodegeneration. Thus there is no surprise that the majority of neurodegenerative diseases have been attributed to an accumulation of oxidative stress. In such a pro-oxidative state, there is progressive damage of cells at the subcellular level eventually leading to cell death through apoptosis (programmed cell death), autophagy or if the damage is severe, necrosis (Scartezzini and Speroni, 2000).

1.1.2.1 OXIDATIVE DAMAGE AS A MAIN CAUSE OF NEURODEGENERATION

Every second of everyday in every cell of the body, a variety of oxygen free radicals collectively known as reactive oxygen species (ROS) are being produced. These little oxidative particles wreak havoc throughout the body by inflicting damage on tissues, cellular molecules and DNA. Oxidative damage is so pinnacle to age-related cognitive decline that even dietary regimes enriched in antioxidants have been associated with some level of protection against neurodegenerative decline (Checkoway et al., 2002; de Rijk et al., 1997; Commenges et al., 2000).

ROS are created as the byproduct of normal biological processes, namely the energy producing process in mitochondria known as oxidative phosphorylation. During this process, electrons are passed between three subunits (I, II and III) of the 'electron transport chain' (ETC), which through a series of biochemical interactions, convert the energy of the electron ions into usable cellular energy. When there is damage or stress upon this delicate chain, the free electrons react erroneously with neutral molecules or become converted by enzymes into other dangerous counterparts. Most commonly, the free electrons will react with the abundantly present and electron-receiving oxygen molecules, hence the name reactive *oxygen* species. In this highly unstable state, oxygen becomes dangerous as it converts into a free radical: an atom, molecule or ion that has unpaired electrons thus existing in a highly unstable form.

ROS invoke damage by interacting with healthy or normal molecules, transferring to them a free electron and making these molecules themselves highly reactive. By far, the most reactive ROS is the hydroxyl radical (OH^-) as it reacts with everything from proteins, to lipids, to DNA to RNA. Ultimately, these molecules will promote damage throughout the cell by propagating a chain reaction of irrevocable damage and eventually cell death.

Age is the prime incident factor for the development of ROS. When we are young, our body still has a grand repertoire of antioxidant mechanisms that are fully functional and constantly protecting our cells against such damage. As we age however, these mechanisms become less and less functional and oxidative damage accumulates. For example, many of the enzymes present to protect the body from oxidants begin to lose functionality such as catalase, glutathione peroxidase and superoxide dismutase. Normally these enzymes take free radicals and metabolize them into harmless products, though when they become less functional with age, free radicals are not metabolized and accumulate proportionally in cells.

Environmental toxins are another main contributor to oxidative stress. Toxins such as herbicides, industrial fumes and synthetic food additives all tip the homeostatic balance of the body towards a pro-oxidative state. Such toxins can works through many mechanisms such as introducing new oxidative particles into the body, inhibiting antioxidant enzymes or invoking mitochondrial damage thus increasing the production of free radicals (Halliwell, 2011).

It is not surprising then to acknowledge that the most important and common link between all neurodegenerative diseases is the induction of a pro-oxidant state. As it was mentioned, the incorporation of antioxidants into one's diet is enough to reduce the risk of developing neurodegenerative disease. But, what if we go further with this idea? Why not incorporate the most powerful antioxidants into the diet and lifestyle routines to greatly reduce the risk of developing any sort of disease? Indeed, herbal products and medicines outlined by traditional medicines allow just this. Many of the herbal medicines utilized in alternative medical techniques, including Ayurveda, are rich in polyphenols: chemical structures resembling a ring that are known to scavenger those damaging free radicals. Hence, the fight against neurodegeneration can start (and end) with the educated intake of Ayurvedic herbal medicines aimed to protect neurons against age- and environmentally-provoked oxidative damage.

Below is a description of the three most devastating neurodegenerative diseases, Alzheimer's disease, Parkinson's disease and Huntington's disease, and how oxidative damage plays a major role in each of their pathologies.

1.1.2.2 ALZHEIMER'S DISEASE

Alzheimer's disease (AD) is the most prevalent neurodegenerative disease facing society today and its main symptoms are progressive short-term memory loss and dementia, which eventually lead to death. Presently, it is estimated that 10% of the population over 60 and 50% over 80 are suffering from AD (Potter, 2013). In light of the rapidly ageing population, the number of cases is expected to quadruple by 2050. The symptoms of AD are severe, beginning with memory loss for recent events and escalating to confusion, aggression, irritability and retroactive memory loss (i.e., forgetting your friends, family and childhood events). Ultimately, the body begins to 'forget' how to preform autonomic functions as the motor systems begin to fail. The heart will stop beating, brain functions will be lost and eventually the patient will die. The severity of the debility and the looming increase in AD prevalence demands preventative measures for the global population.

The cause of AD is unknown. A small fraction of AD cases are familial or early-onset (10–15%) (Alzheimer's Association) and have been associated with mutations in amyloid precursor protein (APP), presenilin 1 and presenilin 2 (Sehgai et al., 2012). The majority of AD cases is sporadic in nature and are correlated with inheritance of the APO?4 allele, an inefficient form of APP. In all of these cases, the improper functioning of APP results in the inefficient breakdown of the beta amyloid protein, thus precipitating of the toxic version of amyloid beta (Aβ) protein, which is more prone to form the Aβ plaques characteristic of AD (Hardy and Higgins, 1992). In addition, there is a high prevalence of aggregated hyperphosphorylated tau proteins known as neurofibrillary tangles or senile plaques (Thompson and Vinters, 2012). Through not fully understood mechanisms, the accumulation of these plaques and tangles leads to the progressive and irreversible loss of neurons, mainly in the cortex and hippocampus (Tundo et al., 2012). It is believes that through the induction of apoptosis, activation of transcription factors and mitochondrial damage, these plaques induce cell death (Ramasamy et al., 2006).

There is another hypothesis regarding the etiology of AD, and this involves the loss of cholinergic neurons. Indeed, one of the earliest neurochemical signs of AD is the loss of cholinergic neurons (Bartus et al., 1982). Indeed, the depletion of acetylcholine in the brain is proposed to have a prominent role in the memory impairments of AD patients (Lahiri et al., 2004; White and Rusje, 2002). Although the evidence of this effect has become controversial, most of the drugs used to treat the symptoms of AD target the deficiencies in the cholinergic system.

Oxidative stress is also a risk factor for the development of AD (Butterfield et al., 2004) and studies have indicated clear markers of oxidative damage in AD brains (Pratico et al., 2000). Indeed increases in lipid peroxidation, protein oxidation and DNA oxidation have all been reported in AD patients and as shown in animal models, induction of antioxidantive mechanisms attenuates AD's progression (Zandi et al., 2004). Some believe that the mechanism of Aβ-induced cell death even may be aggravated by oxidative pathways (Cai et al., 2011). Actually, there is a direct correlation between the concentration of the hydroxyl ion and the co-localization of the hyperphosphorylated form of tau (Takeda et al., 2000). Further, oxidative damage plays a role in cytoskeletal abnormalities, a major pathological feature in AD (Smith et al., 1995). Clearly, many of the pathological pathways involved in the development of AD are aggravated by oxidative stress and this very observation indicates strongly a role for herbal medicines with their potent antioxidant activity for the treatment of AD.

1.1.2.3 PARKINSON'S DISEASE

Parkinson's disease (PD) does not discriminate between race, religion or culture and affects 2% of people over 60 and 5% over 80 worldwide (Lees et al., 2009). It is characterized by the progressive degeneration of dopaminergic neurons in the *substantia nigra*, a midbrain region. The atrophy of these neurons is thought to contribute to the progressive motor deficits characteristic of PD. These symptoms include progressive shaking, slowness in movement, rigidity, depression, dementia and digestive difficulties. PD is a multifactorial disease instigated by both environmental and genetic factors. The etiology of PD remains debatable, though it is believed that the increase in oxidant stress in the brain from aging provokes the progressive degeneration of dopaminergic neurons. This effect can be aggravated by genetic factors, damaged mitochondria and the accumulation of ?-synuclein or Lewy bodies in corresponding brain regions.

Much research has been conducted to discern the precise mechanisms underlying PD, though the causes are diverse and still debated. The majority of PD cases are sporadic with a small percentage being due to genetic factors (1–5%). Several environmental toxins have been identified that increase the risk of PD development, namely the commonly used pesticides and herbicides such as paraquat, rotenone and MPTP. Nevertheless, both the environmental and genetic causes of PD have been found to act through similar mechanisms, the major one being oxidative stress (Tsang and Chung, 2009). Notably, the overproduction of ROS that has been noted in PD patients has been deemed to be a major cause of dopaminergic degradation in PD patients (Danielson and Andersen, 2008). Mitochondrial damage and the accumulation of the aggregated ?-synuclein protein are also important considerations for the degradation of dopaminergic neurons. For example, environmental toxins such as MPTP, paraquat and rotenone all inhibit complexes in the ETC leading to mitochondrial damage. Such damage increases the production of ROS as free electrons being shuttled through the ETC erroneously and at a higher frequency react with biological molecules.

The genetic and environmental risk factors not only increase the production of ROS, but also decrease the processivity of antioxidant enzymes such as superoxide dismutase, catalase and glutathione peroxidase. Finally, the peroxidation of polyunsaturated fatty acids in the neuronal cell membranes (lipid peroxidation) rears a significant source of oxidative stress in models of PD. This initial oxidative damage begins a chain reaction creating many downstream degradation products that are harmful to the neurons (Dexter et al., 1989). Already the antioxidant defenses in neurons is compromised due to their weak defenses, so the further depletion of their potency severely lessens the ability of cells to defend themselves against the onslaught of oxidative stressors.

1.1.2.4 HUNTINGTON'S DISEASE

The pathogenesis of Huntington's disease (HD) is slightly different from AD and PD. HD affects 4 – 10 in 100 000 persons, primarily of Caucasian origin (Sandhir et al., 2013). HD is characterized by a spectrum of cognitive, psychiatric and motor symptoms such as chorea (jerky, uncontrollable movements) and poor coordination. In more advanced stages, HD patients will experience dementia and further complications of the motor deficits namely pneumonia and heart disease. Normally, these symptoms will begin later in life, between 35 and 44 years of age and due to the progression of symptoms, death will ensue about 20 years after symptoms begin. As there is largely a motor deficit and atrophy in the striatum of the post-mortem brains, this area has become the focus of the disease's pathology (Walker et al., 2010).

It is known that the mitochondria play a large role pathogenesis of this disease. Indeed, 3-nitropropoonic acid (3-NP) serves as a good model of HD as it is a neurotoxin that induces a loss of mitochondrial function by selectively inhibiting the activity of Complex II of the ETC (Gould et al., 1995). Similar then to AD and PS, this toxin increases the ROS production in the striatum eliciting HD-like symptoms (Bogdanov et al., 1998).

HD is caused by an autosomal dominent mutation in the Huntingtin gene, namely a series of 'CAG' triplet repeats. CAG encodes for the amino acid glutamine and when the protein of this gene is created, the extra glutamine residues invoke a dominent pathology in a number of ways. The number of CAG repeats in a sufferer of PD varies between individuals, though generally, the more repeats there are, the more severe the disease and the earlier that symptoms will appear. The multiple repeats will erect misfoldings in the Huntingtin protein resulting in toxicity by altering subcellular localization and instigating abnormal interactions with other cellular proteins (Bantubungi and Blum, 2007). Although the precise function of the Huntington gene in humans remains unclear, it is known to interact with at least 100 target genes and is involved in a diverse array of biological functions (Goehler et al., 2009). Overall, excitotoxicity, dopamine toxicity, metabolic impairment, mitochondrial dysfunction, oxidative stress, apoptosis and autophagy have all been implicated in the progressive degeneration observed in HD (Gil and Rego, 2008). What is known is that the Huntingtin protein is involved in the regulation of apoptosis (cell death) pathways. The mutated form of Huntingtin is postulated to activate caspases through damage inflicted on the ubiquitin-protease system. These caspases go further to stimulate pathways eventually leading to controlled cell death or apoptosis.

1.1.3 HERBAL TREATMENTS OF NEURODEGENERATIVE DISEASE

As clearly demonstrated in the above descriptions of the neurodegenerative diseases, accumulation of oxidative stress is the primary and common risk factor for the development of age-related cognitive decline. Thus, by actively countering the accumulation and formulation of such harmful species, neurodegeneration can be delayed, potentially indefinitely.

In this comes the power of herbal medicines. As mentioned above, the majority of herbal plants are a rich source of complex polyphenols, which is a class of chemical structures characterized by a series of interconnected benzyl rings. These structures are especially efficient at absorbing free electrons and hence eliminating free radicals floating erroneously in the cell. Some of the most important polyphenols are the tannins popularly found in teas and wine and a large variety of plants and fruits such as pomegranates, berries, grapes, dark-colored legumes and of course, chocolate. Thought to be anti-nutritive due to their protein-scavenging activities, they are indeed powerful antioxidants and have many beneficial effects on physiology.

Flavonoids are a class of polyphenols, which are the most potent natural free radical scavengers. There are many classes of flavonoids corresponding to their structure and function. Though overall, flavonoids are beneficial to every aspect of physiology and have been reported to treat cancer, improve allergies, boost the immune system, have anti-aging, anti-bacterial, anti-fungal, anti-inflammatory properties and notably, prevent the onset of neurodegenerative diseases with their antioxidant activity.

There is another category of herbal constituents that convey a broad host of medicinal properties known as alkaloids. Alkaloids are a diverse group of low molecular weight nitrogenous compounds that have been used for, and against humans for thousands of years. One thing that binds this group is the presence of nitrogen and their characteristic bitter taste. Alkaloids are potent compounds found in plants that naturally give them the ability to fight pathogens and predators alike. They can be poisonous as well as medicinal. Many of the most common medicines are derived from plants including morphine (pain-killer), caffeine (stimulant), Quinine (anti-malaria) and even nicotine. Although there is no solid definition of what as alkaloid actually is, the majority of active alkaloids used in medicine contain multiple ring structures similar to the polyphenols. This would then suggest that in addition to their specific activities that each have some extent of oxidative scavenging activity.

In this comes the power of herbal medicines. Herbs used in Ayurveda are a rich source of diverse polyphenols, flavonoids and alkaloids thus providing a broad array of beneficial physiological effects. In popularized allopathic medicine, isolated treatments are normally administered to patients with complex diseases, in the hope that alleviating the ill in one biological pathway will ultimately alleviate all the symptoms and complications of the disease. In holistic and specifically Ayurvedic practices, treatments are not focused on correcting an isolated pathway, but rather to broadly affect many pathways to ultimately reinstate physiological homeostasis. It is only when the entire body exists in a state of health that balance of the *doshas* can be reinstated and one's *vikriti* eliminated. It is this kind of diverse action from herbal medicines that is required to prevent and treat the multi-functional pathology of neurodegenerative diseases. Below, a selection of powerful Ayurvedic herbs will be described in terms of their chemical content and the traditional and scientific evidence supporting their role in neurodegenerative diseases.

1.2 WITHANIA SOMNIFERA

Withania somnifera is commonly known in Sanskrit as *Ashwagandha*, which literally means 'the smell of a horse'. This characteristic was ascribed to the freshly harvested roots of the plant, which have the potent essence of horse urine. This herb is also known as Winter cherry in English or Indian ginseng as it contains many of properties of the popularized and highly potent Asian ginseng (*Panax ginseng*) used in Chinese medicine. *Withania somnifera* belongs to the Solanaceae family and is normally harvested in India, Africa and the Mediterranean. It is a small shrub-like plant growing to approximately 1.5 m tall. It is related to the tomato plant in that it has small yellow flowers that turn into small red fruits at maturity.

1.2.1 WITHANIA SOMNIFERA IN AYURVEDA

Withania somnifera can boldly claim to be one of the most revered herbs in Ayurvedic medicine as it has been used since the dawn of its existence over 5000 years ago and is highly cited in the ancient texts (Charak Samhita, 1949). Throughout the tradition of Ayurveda, it has been shown to have a variety of medicinal properties including tonifying, aphrodisiac, narcotic inducing, diuretic, anthelmintic, astringent, thermogenic and stimulating. *Withania somnifera* has a range of *rasa*'s (tastes) namely *tikta* (bitter), *katu* (pungent) and *madhura* (sweet) with a *laghu* (light) and *snigdha* (oily) *guna* (quality). It has an *ushna* (warm) *virya* and *madhura* (sweet) *vipaka*. It is best received to pacify *vata* and *kapha* types, though can be safely used by *pitta* types unless taken in excess where the excess heat will be unbalancing.

Of its many contributions to health, *Withania somnifera* is best known as a *medhya rasayan*, literally a mental tonic. In children, *Withania somnifera* has been shown to stimulate mental and intellectual capacity while in aged persons, can relieve debility and promote diminished memory faculties (Sharma et al., 1999). In many modern studies, *Withania somnifera* was shown to enhance the function of the brain, improve memory and act as an overall tonic to the central nervous system (Singh et al., 2011). However the consumption, however the delivery, *Withania somnifera* is a trusted neuroprotective agent in Ayurveda that prevents age-related neuronal degeneration, allows the body to adapt to incoming stresses (adaptogenic), promotes neural growth and provides a diverse range of antioxidant protection to the central nervous system (Battacharya et al., 2001; Bhattacharya et al., 1999; Gupta et al., 2003; Mishra et al., 2000).

Traditionally, the roots of *Withania somnifera* are consumed as a fine powder mixed with honey, water and/or ghee, as a *churna*; though, all parts of this plant have medicinal properties. The leaves are bitter in taste and are recommended more for the treatment of fever; the flowers are astringent and act more as depurative, diuretic and aphrodisiac; and the seeds are anthelmintic, meaning that they dispel parasitic worms. Despite the focus of this herb's potency being in the roots, recent studies have shown that the leaf extract too has promising, or even better potential to act as a neuroprotective agent (Konar et al., 2011). Interestingly, extensive studies have shown that there is no toxicity of this herb, at any dose, at any age or in any condition (Mishra et al., 2000).

1.2.2 CHEMICAL COMPOSITION

The vast arrays of actions of Withania somnifera are supported by an underlying complexity of active components. *Withania somnifera* contains various alkaloids, steroidal lactones and saponins (Mishra et al., 2000) with the major active ingredient being the glycowithanolides, a subtype of steroidal lactones that provide most of its antioxidant properties. Of particular importance to neurodegenerative diseases are the withanolides, withanone and withaferin. These make up the bulk of the neuroprotective agents, also conveying potent antioxidant functions. Of particular note, the action of *Withania somnifera* is dependent on the method of extraction. In one study, a water extract of *Withania somnifera* leaves (but not alcoholic extract) was found to selectively kill cancer cells, while leaving normal cells intact. Interestingly, this was found to be from the unique balance of withanone and withaferin A in the water extract. If the concentration of withaferin A is too high, them the cancer cells and normal cells will succumb to apoptosis and only in the water extract of the leaves is the perfect harmony found (Widodo et al., 2010).

1.2.3 ANTIOXIDANT ACTION OF WITHANIA SOMNIFERA

As mentioned, *Withania somnifera* has a potent and diverse battery of anti-oxidative agents and it is in these constituents, namely the phenolic compounds, that *Withania somnifera* gets its power to fight neurodegenerative diseases (Prakash et al., 2007). In particular, isolated glycowithanolides protect neurons against lipid peroxidation (Bhattacharya et al., 2000; Chaurasia et al., 2000), extracts of withanolides and sitoindosides (VII–X) enhance catalase and glutathione peroxidase activities *in vivo* (Chaurasia et al., 2000; Bhattacharya et al., 2001; Bhattacharya et al., 1997) and *Withania somnifera* extracts can accentuate glutathione concentrations by increasing glutathione peroxidase activity to ultimately inhibit the formation of TBARS, another oxidative marker (Prakash et al., 2013).

Moving to *in vivo* studies, it was shown in an induced oxidative-stress model of neurodegeneration (streptozotocin) that *Withania somnifera* can completely protect against the behavioral phenotypes, namely memory loss and can maintain the level of antioxidant enzymes while suppressing the accumulation of oxidative particles (Ahmad et al., 2013). Interestingly, *Withania somnifera* also repressed the production of nitric oxide (NO), a potent oxidant, during a chronic stress regime (Bhatanagar et al., 2009). What is interesting about this observation is that *Withania somnifera* appeared to act by suppressing corticosterone release and the consequent activation of acetylcholinesterase, thus increasing serotonin levels in the hippocampus and indirectly inhibiting NO production. This indirect antioxidant potential indicates that *Withania somnifera* could additionally act as an anti-AD agent by upregulating cholinergic signaling in the brain.

1.2.4 THE NEUROREGENERATIVE ABILITY OF *WITHANIA SOMNIFERA*

In modern science, it is of the understanding that adult neurons cannot regenerate; however, *Withania somnifera* root extracts have been shown in numerous studies to stimulate the regeneration of neurons and encourage dendritic arborization (Kuboyama et al., 2002; Tohda et al., 2005). Indeed, the methanolic extract of *Withania somnifera* stimulated the increase in dendritic markers in human neuroblastoma cells (Kulkarni et al., 1993) and using these cells, *Withania somnifera* extracts together with a selection of withanolide derivatives were shown to promote dendrite outgrowth (Tohda et al., 2000). Extending this into animals, withanolide A, withanoside IV and withanoside VI were shown to restore pre- and post-synapses in rodents lost to the injection of Aβ peptide, a neurodegenerative aggregative protein (Kuboyama et al., 2005). Although the regenerative potential has not yet been proven in neurodegenerative disease, the likelihood that *Withania somnifera* could be curative in diseases with neurodegeneration is promising.

1.2.5 WITHANIA SOMNIFERA IN ALZHEIMER'S DISEASE

The major impairments in AD are ultimately due to neuronal atrophy and synaptic loss (Dickon and Vicker, 2001), though the pathological mechanisms behind these events still remain elusive. One popular hypothesis states that loss of cholinergic signaling in the hippocampus instigates the selective neuronal loss in AD patients and this effect may be precipitated by the accumulation of Aβ and neurofibrillary tangles. Nevertheless, *Withania somnifera* has been boldly acclaimed to slow, stop, reverse and remove neuritic atrophy and synaptic loss in AD models (Singh et al., 2011).

In animal models of AD, *Withania somnifera* extracts have been shown to reverse typical behavioral deficits such as memory loss and poor cognitive functioning (Bhattacharya et al., 1995). Indeed, *Withania somnifera* root extract and isolated withanolides have been shown to inhibit acetylcholinesterases, the enzymes responsible for degrading acetylcholine, both *in vitro* and *in vivo* (Choudhary et al., 2005). Further, in an *in vivo* AD transgenic model, *Withania somnifera* was shown to increase the level of acetylcholine in the hippocampus by inhibiting acetylcholinestase (Ahmad et al., 2013). With this evidence, it may be postulated that *Withania somnifera* imparts its beneficial effects on the memory deficits of AD patients by reconstituting the levels of acetylcholine.

The regenerative potential of *Withania somnifera* extracts was demonstrated in an animal model of AD instigated by injections of Aβ. In this study, chronic oral administration of withanoside IV rescued the memory loss phenotype, attenuated the axonal, dendritic and synaptic losses and stimulated synaptogenesis, axonal growth and dendritic arborization (Kuboyama et al., 2006). This is a powerful study as it indicates that components of *Withania somnifera* can work *in vivo* to therapeutically cure neurodegenerative disease and re-grow damaged neurons.

Another interesting action of *Withania somnifera* against AD was recently defined, not in the central nervous system, but in the periphery.

There is a degrading protease in the periphery, namely in the liver, called neprilysin that can potently destroy Aβ plaques. *Withania somnifera* was found not only to upregulate the processivity of neprilysin, but to also enhance the trafficking of Aβ plaques out of the brain via the lipoprotein receptor-related protein. Hence, *Withania somnifera* can be used to directly reduce the number of Aβ plaques and thus pharmacologically eliminate AD (Sehgai et al., 2012).

From these diverse studies, it is no wonder that *Withania somnifera* is regarded as the wonder herb of Ayurveda. It is able to battle the oxidative stress behind AD's pathology, actively remove and degrade plaques and then build back the lost neurons. It is a wonder why, after all this evidence, that *Withania somnifera* has not been developed further in the western world as a canonical treatment of AD.

1.2.6 *WITHANIA SOMNIFERA* IN PARKINSON'S DISEASE

The pathology of PD is quite different from AD, though similarly, oxidative stress is a potent risk factor for its symptomatic development and specific dopaminergic cell loss. PD's symptoms are primarily characterized by motor deficits such as resting tremor and rigidity and *Withania somnifera* has been shown to effectively reduce these deficits in a toxin (MPTP + paraquat)-induced mouse model (Prakash et al., 2013). *Withania somnifera* root extracts have also been found to improve essentially all aspects and risk factors of PD from increasing the levels of dopamine metabolites (RajaSankar et al., 2009), to increasing dopaminergic D2 receptor binding (Ahmad et al., 2005), to even improving the secondary effects of PD such as general mental health and brain aging (Singh et al., 2008). In particular, *Withania somnifera* displays potent antioxidant potential by enhancing antioxidants and antioxidant enzymatic activities in the striatum (glutathione, catalase, superoxide dismutase and glutathione peroxidase) and decreasing oxidative stress markers (ROS, nitric oxide, lipid peroxidation) in mostly all models of PD from 6-OHDA (Nagashyana et al., 2000; Ahmad et al., 2005) to MPTP (RajaSankar et al., 2009) to rotenone (Manjunath et al., 2013) and a combined progressive model of PD (Prakash et al., 2013).

As in AD, the diverse functionality of *Withania somnifera* has been extensively shown to rescue phenotypes, biochemical changes and neuronal degradation in the PD model thus giving it a keen outlook towards its therapeutic potential.

1.2.7 *WITHANIA SOMNIFERA* IN HUNTINGTON'S DISEASE

HD has a slightly different pathology than the other canonical neurodegenerative diseases such as AD and PD, though the breadth of *Withania somnifera*'s action even allows it to be beneficial in the treatment of HD. 3-NPA is toxin produced by a variety of fungi and sugarcane found commonly in food. It is an irreversible inhibitor of the ETC complex II thus acutely increases oxidative damage and is potently toxic to humans. Due to this effect, it has been linked to lesions that are similar to those found in HD and can used as a suitable oxidative model of this disease.

In the 3-NP model, the chronic treatment of *Withania somnifera* enhanced the memory performance and significantly restored several antioxidant pathways in the body and acetylcholinesterase levels in several regions of the brain (Kumar et al., 2008). In a similar study, chronic *Withania somnifera* treatment rescued behavioral abnormalities induced by 3-NP (gait abnormalities) and restored the biochemical changes in this model (Kumar et al., 2009). This effect was dose- and time-dependent indicating the direct potency of *Withania somnifera*'s antioxidant effects against HD.

In conclusion, *Withania somnifera* is a potent herb whose antioxidant activity has been characterized to elicit neuroprotective effects against neurodegenerative diseases. There is indirect evidence too linking the other actions of *Withania somnifera* to various neuroprotective pathways, yet these have not been fully investigated. Finally, the neuroregenerative potential of *Withania somnifera* makes it an attractive therapeutic as it does not only prevent damage from occurring, but can heal damaged neurons and reconstitute a damaged brain.

1.3 BACOPA MONNIERI

Bacopa monnieri (L.) is commonly referred to by its Sanskrit name *Brahmi,* though this creates a slight misconception as the name *Brahmi* is shared with a similar, yet distinct plant, *Centella asiatica*. The name *Brahmi* is derived from the name of the divine God *Brahma*, the 'supreme creator' who worshipped throughout India. This name is appropriate as *Bacopa monnieri* improves overall brain function and stimulates creative activities. It is in these properties that the clarity of mind is established to improve the connection to the unconscious universe, or *Brahma*. *Bacopa monnieri* does not only stimulate the mind, but the higher connections to the ultimate supreme universe. It must be mentioned that this property of *Bacopa monnieri* perfectly outlines the holistic nature of Ayurveda medicine. The physical medicines available to treat any disorder is just one of the powers which they offer. Every step one takes to closer achieve perfect health actually is bringing them closer to spiritual enlightenment.

Barcopa monnieri is a bitter tasting plant found primarily in wetlands and muddy shores in India, Sri Lanka, China and Pakistan. It is a small, branched creeping herb with small leaves and white/light purple flowers belonging to the family Schropulariacae. Its taste (*rasa*) is described as *madhura* (sweet)*, tikta* (bitter) and *kasaya* (astringent), with a *laghu* (light) and *sara* (mobile) *guna* (quality). It has a *sita* (cold) *virya* and *madhua* (sweet) *vipaka*. *Bacopa monnieri* is balancing to all *doshas* though due to the lightness and mobility of this herb, excessive intake can be unbalancing to *vata* individuals.

In 1931, the main alkaloid present in *Barcopa monnieri* was identified as 'brahmine' (Russo et al., 2005). After some time, the alkaloids were specifically identified to be nicotine and herpestine (Chopra et al., 1956). Ultimately, *Barcopa monnieri* was found to be a pharmacologically complex mixture of many structurally related compounds (Murthy et al., 2006). The main active chemical constituents of *Barcopa monnieri* were found to be a collection of steroidal saponins known as bacosides, principally bacoside A and B (Deepak et al., 2004).

Barcopa monnieri has been used as a nervine tonic for millennia in Ayurveda and also falls under the category of a *medhya rasayan*, a cognitive enhancer and anti-anxiety agent. Importantly, *Barcopa monnieri* can work both as a palliative treatment for age-related memory decline and for the enhancement of mental acuity in young and middle-aged people alike (Sharma et al., 1987; Roodenrys et al., 2002). This diversity of function is interesting as *Bacopa monnieri* is able to guide the modeling of neurons in developing brains and also enhance neuronal networks in aged, undividing brains. Although therapeutically *Barcopa monnieri* is most well known for treating depression (Sairam et al., 2002) and rescuing age-related decline in cognition, it also potently reduces anxiety and senility, acts as an adaptogen and boosts immunity. In modern understanding, *Barcopa monnieri* is recognized as having neuroprotective and nootropic activities (Dave et al., 1993) and has henceforth become recognized for its potential uses against neurodegenerative diseases. Most importantly, *Bacopa monnieri* is able to enhance memory and cognitive function without any known side-effects (Calabrese et al., 2008).

Barcopa monnieri elicits its effects through many diverse mechanisms. Notably, *Barcopa monnieri* has potent ROS scavenging activity (Mishra et al., 2011) and is effective in combating oxidative stress (Bhattacharya et al., 2000; Russo et al., 2003). In both aged and middle-aged animals, bacosides can reduce the accumulation of oxidative stress markers including lipid peroxidation and nitric oxide. Interestingly, bacosides actively increased the activity of select antioxidant enzymes such as glutathione reductase, though neither superoxide dismutase nor catalase (Rastogi et al., 2012). Further, in an *in vivo* system of oxidative damage elicited by acrylamide treatment, *Bacopa monnieri* leaf powder in both mice and *Drosophila* effectively prevented oxidative injury and associated cell death (Shinomol et al., 2013).

1.3.1 ROLES OF *BACOPA MONNIERI* IN NEURODEGENERATION

As previously demonstrated, increasing the antioxidant potential of any animal is a beneficial treatment of neurodegenerative disease. Indeed rats treated with *Bacopa monnieri* were shown to have reduced levels of aging biomarkers, namely lipofuscin. *Bacopa monnieri* is postulated to initiate the majority of its effects through its antioxidant potential (Russo et al., 2003; Bhattacharya et al., 2000). Indeed, *Bacopa monnieri* was shown to have abundant antioxidant effects in various areas related to memory in the brain including the hippocampus, frontal cortex and striatum (Bhattacharya et al., 2000). Finally, *Bacopa monnieri* was also shown to reduce the oxidative effects in the brain against those smoking cigarettes, showing how *Bacopa monnieri* ubiquitously benefits the entirety of physiology (Anbarasi et al., 2005).

1.3.2 BACOPA MONNIERI IN ALZHEIMER'S DISEASE

Barcopa monnieri has also been shown to be protective against the ageing and disease markers of various Senile Dementia of Alzheimer's type (SDAT) diseases. In primary cortical neurons, it was shown that *Barcopa monnieri* could prevent A? induced cell death, likely through antioxidant mechanisms (Limpeanchob et al., 2008). Further, *Bacopa monnieri* was found to reverse the memory impairments in an *in vivo* model of AD (colchicine) that is defined by specific cellular damage, increased oxidative damage and depletion of choline acetyl-transferase activity. In addition, this study showed that *Bacopa monnieri* could restore the activity of membrane bound enzymes that were altered in colchicine treatment, namely Na/K ATPase and acetylcholinesterase (Saini et al., 2012).

Going further, in a transgenic mouse model of AD, the antioxidant potential of *Bacopa monnieri* was shown to suppress the accumulation of Aβ levels in the brain (Dhanasekaran et al., 2004). In a specific model of AD in mice (ethylcholine aziridnium ion), *Bacopa monnieri* rescued the behavioral memory deficits and potentiated cognitive enhancement evidenced through the mitigation of cholinergic neuronal densities (Uabundit et al., 2010). Indeed, it seems that *Bacopa monnieri* is able to potentiate the cholinergic system as indirectly evidenced in the above *in vivo* studies. To support this, there has been a direct *in vivo* study indicating that *Bacopa monnieri* can modulate the cholinergic system and increase the level of acetylcholine in rat brains (Bhattacharya et al., 1997). This effect is thought to be correlated with enhanced memory and rescuing of motility in SDAT disorders.

Altogether, it seems that *Bacopa monnieri* has the potential to protect against AD through numerous multifactorial mechanisms. *Bacopa monnieri* offers both oxidative protection and protection against acetylcholine loss in the brain, effects that are reflected in the betterment of biochemical and symptomatic behaviors associated with AD such as loss of memory and motor coordination.

1.3.3 BACOPA MONNIERI IN PARKINSON'S DISEASE

There is a large body of evidence supporting the role of *Bacopa monnieri* in the protection against the onset of PD. Again, the majority of the actions of *Bacopa monnieri* may be attributed to its anti-oxidative effects. Indeed,

Bacopa monnieri was shown in a toxic-induced PD model to reduce free radicals such as nitric oxide, superoxides and hydroxyl radicals (Hosamani et al., 2010). In particular, pre-treatment of *Bacopa monnieri* extracts in mice was shown to eradicate rotenone-induced oxidative stress and neurological death in the striatum and other brain regions by upregulating antioxidant factors (Hosamani et al., 2009). This effect can be extended into the protection from protein oxidations implicated by protein carbonyl factors (Shinomol et al., 2012). In all of these studies, *Bacopa monnieri* was synergistically able to rescue the behavioral deficits characteristic of PD, namely locomotor abilities, in addition to the biochemical parameters.

However, the action of *Bacopa monnieri* goes deeper than antioxidantive potential. Bacosides, one of the main constituents of *Bacopa monnieri*, fundamentally improve the pathological parameters of PD by preventing the age-related deficits in monoaminergic neurotransmitters including serotonin, dopamine and epinephrine (Rastogi et al., 2012). In a *C. elegans* model of PD (both a genetic and toxic model were used), *Bacopa monnieri* was shown to prevent the aggregation of α-synuclein, degeneration of dopaminergic neurons and to promote longevity in the nematodes (Jadiya et al., 2011). These studies suggest that *Bacopa monnieri* in addition to being an effective stimulator of antioxidant mechanisms in the brain, also fundamentally change cellular mechanisms underlying these deteriorating neurodegenerative diseases.

1.4 CENTELLA ASIATICA

Centella asiatica is a herb well known in both Ayurvedic and Chinese medical literature. In India, *Centella asiatica* is also known as *manukaparni* or in English, Indian pennywort and in the Indonesian islands as *gotukola*, the "miracle elixir of life". Confusingly, *Centella asiatica* also shares its Sanskrit name *Brahmi* with its more popular family member *Bacaopa monnieri*, but it is indeed a different herb altogether.

Centella asiatica (family *Umbellifere*) is a slender creeping plant native to countries like India, Sri Lanka, Madagascar, South Africa and China, growing in moist habitats at altitudes below 1800 m. It has fan-shaped reddish-green leaves poised around small white or light purple flowers and a small fruit. This plant grows in bunches, in an interconnected web of mutually supporting specimens.

In China, *Centella asiatica* was traditionally used as an anti-spasmodic and anti-epileptic herbal treatment as it was revered for its ability to evoke extreme relaxation and mental calmness. In addition, it was used to treat emotional disorders that were thought to originate in the body (Hagemann et al., 1996). It was later found that this calming effect was due to the ability of *Centella asiatica* to temporarily induce the production of GABA, an inhibitory neurotransmitter in the brain implicated with calming the mind. Indeed, recent investigations indicate that animals induced with epilepsy episodes benefited greatly from *Centella asiatica* treatment by reducing the burden on over-loaded cholinergic neurons (Visweswari et al., 2010).

Centella asiatica was introduced later into the repertoire of Ayurvedic herbs but at this time was recognized as the herb of longevity. It was mentioned in the *Sushruta Samhita* as an effective wound healer and in the *Charaka samhita* as an enhancer of intelligence. Though, it was not until the nineteenth century that this herb became more widely used with a regained variety of therapeutic potentials. *Centella asiatica* falls under the "bitter root group" with alternative and tonic properties and is appreciated for its memory boosting and fertility increasing properties. Today, it is regarded to reduce mental fatigue, anxiety and treat headache, body ache, insanity, asthma, leprosy, ulcers and wounds (Gohil et al., 2010). Of particular interest, it has been categorized in the Ayurvedic literature as a stimulatory nervine tonic (*medhya rasayan*) with the ability to stimulate intelligence (Veerendra et al., 2002). This herb has *tikta* (bitter), *kasaya* (astringent) and *madhura* (sweet) *rasa*'s (tastes), a *shita* (cold) *virya* and has a *laghu* (light) and *sara* (mobile) *guna*. After digestion (*vipaka)* it takes on a *madhura* (sweet) quality. The cooling effects of *Centella asiatica* make it especially effective at balancing disorders of *pitta*, though it is balancing for all *doshas*.

This herb's broad range of action can be reflected in its broad range of active components. The major constituent is the battery of polyphenols (Zainol et al., 2003) with the triterpenes (including asiaticoside) being the most biologically active (Inamdar et al., 1996) and primarily responsible for the wound healing effects. *Centella asiatica* also contains large quantities of the triterpenes, madecassic acid and madecassoside (Inamdar et al., 1996; Munduvelil et al., 2010; Zhang et al., 2007). Sharing with its popular brother, *Bacopa monnieri*, *Centella asiatica* also contains the potent brahmoside and brahminoside, agents, which may contribute to the effects it has on the central nervous system.

Centella asiatica is normally consumed as a tea, with the dried leaves being soaked in boiling water and sweetened with honey. The raw leaves can also be used on salads or in curry. All parts of this plant however have been shown to have great benefit, from the seeds, to roots, to leaves to branches.

Many years ago, *Centella asiatica* was recognized as an overall tonic for memory enhancement and general mental ability (Apparao et al., 1973). Indeed, traditionally in Ayurveda, 3–4 leaves of *Centella asiatica* can be mixed with black pepper and given to children in order to enhance memory. Even in mentally retarded children, CA was shown to improve their general mental ability (Kuppurajan et al., 1978). In contrast, healthy yet elderly volunteers responded well to a high dose of the plant extract, self-reporting improved mood and cognitive function thereby indicating a pharmaceutical ability of *Centella asiatica* to treat age-related cognitive decline (Wattanathorn et al., 2008). Such memory-enhancing effects were further reflected in its action as a potent anti-depressive agent (Liu et al., 2004).

1.4.1 CENTELLA ASIATICA IN NEUROTOXICITY

Pharmacological studies have indicated a role for *Centella asiatica* in various aspects of neurotoxicity. Extracts of *Centella asiatica* have been shown to have neuroprotective (Haleagrahara et al, 2010), antioxidant (Gupta et al, 2006), anti-inflammatory (Guo et al., 2004) and anti-genotoxic (Siddique et al., 2008) properties. In the D-galactose induced model of neurotoxicitiy, *Centella asiatica* extract protected against oxidative damage, memory impairment, behavioral alterations and mitochondrial enzyme complex activities (Kumar et al., 2011).

Indeed, like all herbs effective in treating neurodegeneration, *Centella asiatica* has effective antioxidant promoting functions (Kumar et al., 2002). The administration of a crude extract of *Centella asiatica* to cancerous mice increased the action of antioxidant enzymes, namely superoxide dismutate, catalase and glutathione peroxidase (Cesarone et al., 1992; Jayashree et al., 2003). Indeed, oral supplementation with *Centella asiatica* extract for 60 days prevents age-related changes in the antioxidant defense system, lipid peroxidation and protein carbonyl formation in various rat brain regions implicated in cognitive function. Thus, *Centella asiatica* exerted significant neuroprotective effect and proved efficacious in protecting rat brain against age related oxidative damage (Subathra et al., 2005). Finally, in an oxidative-induced impairment model of cognition (streptozotocin), *Centella asiatica* was shown to elevate the level of cognition while rescuing select antioxidant mechanisms such as catalase and glutathione levels (Veerenda et al., 2003). As this model has also been used to study oxidative damage in Alzheimer's disease, this herb has recently been hypothesized to be a reliable treatment against advanced neurodegenerative diseases.

1.4.2 CENTELLA ASIATICA IN ALZHEIMER'S DISEASE

Accordingly to the hints suggesting a role for *Centella asiatica* in neurodegenerative diseases, a study was conducted that showed in an Aβ expressing cell line and in a primary embryonic cortical culture that *Centella asiatica* did indeed reduce the AD pathology in these cells (Xu et al., 2008). Further, in a hippocampal cell line, select asiaticoside derivatives were able to reduce the Aβ induced cell death (Mook-Jung et al., 1999). In a mouse model of Alzheimer's disease, a water extract of *Centella asiatica* lowered the high Aβ burden, however in human neuroblastoma cells, *Centella asiatica* did not show anticholinesterase activity nor protect neurons from oxidative damage or glutamate toxicity (Soumyanath et al., 2012). So, the mechanism of action of *Centella asiatica* seems to be quite different from its brother, *Bacopa monnieri* who clearly has some effect on the cholinergic signaling in the pathology of Alzheimer's. *Centella asiatica* on the other hand seems to target more the formation of Aβ plaques, though the mechanism of its neuroprotective effects remains to be characterized in mammalian models. This suggests that perhaps it is not only the common constituents with *Bacopa monnieri* that are eliciting the protective effects of *Centella asiatica*.

1.4.3 CENTELLA ASIATICA IN PARKINSON'S DISEASE

Centella asiatica too has been shown to have beneficial effects on the pathology of PD. In the MPTP model of PD, both madecassoside (Xu et al., 2013) and asiaticoside (Xu et al., 2011) were shown to be effective at antagonizing locomotor dysfunction and associated neurotoxicity. In particular, these agents rescued dopamine levels in the striatum, increased the anti-oxidantive potential and induced an anti-apoptotic effect by increasing the expression of anti-apoptotic factors. Thus, similar to many other herbal products, to anti-oxidative potential of *Centella asiatica* seems to rear a positive effect of PD prognosis.

1.4.4 CENTELLA ASIATICA IN HUNTINGTON'S DISEASE

Finally, the antioxidant potency of *Centella asiatica* also makes in an effective treatment against Huntington's disease. It was shown in prepubertal mice that treatment with *Centella asiatica* extract decreased the oxidative stress markers induced by 3-NP in multiple brain regions (cortex, cerebellum, hippocampus and striatum) including lipid peroxidation, ROS generation, hydroperoxide generation and protein carbonyl formation. In particular, a reduction of oxidative damage in the mitochondria was noted (Shinomol et al., 2008a, b, 2010).

1.4.5 CENTELLA ASIATICA IN DENDRITIC OUTGROWTH

In addition to neuronal protection, *Centella asiatica* can also positively alter neuronal morphology by stimulating dendritic outgrowth. Leaf extracts from *Centella asiatica* were shown to stimulate neuronal dendritic growth in hippocampal CA3 neurons (Mohandas et al., 2006; Rao et al., 2005). Further, in rats, 6 weeks of treatment with *Centella asiatica* extract lead to a significant increase in the length and branching points of dendritic spines (Shinomol et al., 2008a, b, 2010). A similar study showed that rats treated with fresh *Centella asiatica* leaf extract for long periods had an increase in the dendritic length (intersections) and dendritic branching points (Mohandas Rao et al., 2006). Finally, CA was shown to contribute to the activity of BDNF thereby supporting the neuro-regenerative properties. Indeed the description of this last set of effects is important for the development of *Centella asiatica* as a therapeutic agent, as it has the potential to reestablish damaged neurons while protecting them from further damage from oxidative stressors.

1.5 JATAMANSI

Jatamansi is the gem of Ayurvedic nervine tonics. Its botanical name is *Nardostachys jatamansi* but it is more commonly known just as *jatamansi* or in Hindi as *balchara* or in English, Nardus root or spikenard. This name comes from the appearance of the plant's root where '*jata'* means dreadlock and '*mansi'* meaning human. Indeed the roots are a long thread-like structure resembling dreadlocks in human hair.

Jatamansi belongs to the *Valerian* family and grows namely in the Himalayas of Nepal, China and India at altitudes of 3000–5000 m. The characteristic appearance of *jatamansi* is its attractive pink bell-shaped flowers. This plant grows to about 1 meter in height and is harvested year-round.

In the earliest time, the rhizosomes of *Jatamansi* were crushed and distilled to make a highly viscous oil used for religious ceremonies, as perfumes and as a medicine to treat insomnia, birth defects and other minor ailments. This plant has a particular importance in historical Judaism where it was used to represent Saint Joseph in Catholic iconography. Indeed there are several references to the use of this herb in Judaic ceremonies in the Hebrew Bible namely as a special incense.

In Ayurveda, *jatamansi* is namely used to treat various neurological diseases such as epilepsy, convulsions, mental weakness, anxiety, depression and hysteria (Uniyal et al., 1969). With more recent advancements, *jatamansi* has been incorporated into various treatments for PD and AD.

Ayurveda described *jatamansi* as having three *rasa's* (tastes), *tikta* (bitter), *kasaya* (astringent) and *madhura* (sweet). It has a *laghu* (light) guna, *sita* (cold) *virya* and *katu* (pungent) *vipaka. Jatamansi* is especially useful at battling *vata* imbalances, the main *dosha* responsible for diseases of the brain. Traditionally, *jatamansi* is used to improve intellect, treat depression and even calm insanity. It even has positive effects on mood disorders such as insomnia, aggressiveness and stubbornness.

Jatamansi is most renowned in Ayurveda to act as an enhancer of learning and memory, a nootropic agent. Indeed this herb has been shown to improve the memory of young rodents and even ameliorate the decline in memory of aging rodents (Joshi et al., 2006). In particular, *jatamansi* was shown to protect the hippocampus from the debilitating effects of chronic stress, indicated by the behavioral rescue of various memory tasks (Karkada et al., 2012).

Constitutionally, *jatamansi*'s potency is derived from its volatile oils, two alkaloids and an acid adeptly named jatamansic acid. Chemically, the components include sesquiterpenes and coumarins with the active ingredients Jatamansone or valeranone (Chatterji et al., 1997; Bagchi et al., 1991; Rueker et al., 1993).

The seeds and roots of *jatamansi* are typically used in medical preparations. Ethanolic extracts of the roots have been the focus of much of the modern research of this herb and have been shown to have anticonvulsant activity (Rao et al., 2005), antidepressant activity (Dhingra et al., 2008) and neuroprotective effects (Salim et al., 2003).

1.5.1 ANTIOXIDANT PROPERTIES OF JATAMANSI

The antioxidant capabilities of a hydroalcoholic extract of *jatamansi* were first assessed *in vitro* and with a large battery of antioxidant tests, it was determined that *jatamansi* has a very potent antioxidant abilities, in particular for scavenging superoxide anions as well as nitric oxide (Sharma et al., 2012). Indeed, *Jatamansi* has a strong antioxidant effect, comparable to well-known natural antioxidants (Dugaheh et al., 2013). In an *in vivo* model of catalepsy (muscular rigidity and fixity of posture; a common symptom of PD) *jatamansi* was able to eliminate the physical symptoms by enhancing several anti-oxidative enzymes, namely catalase, superoxide dismutase and glutathione peroxidase (Rasheed et al., 2010). In the brain utilizing a stress-induced model of neurodegeneration, the hydroalcoholic extract of *jatamansi* specifically provided antioxidant protection in both the brain and in the periphery (Lyle et al., 2009).

1.5.2 JATAMANSI IN ALZHEIMER'S DISEASE

Jatamansi has been shown by a battery of studies to positively affect cognition and improve mental symptoms of AD. In mice, *jatamansi* was shown to have a powerful nootropic effect. Administration of an ethanolic extract of *jatamansi* significantly improved the learning and memory of young mice. Further, it also reduced age-related amnesia, possibly due to the facilitations of cholinergic transmission in the brain and its potent antioxidant potential (Joshi et al., 2006).

Notably, *jatamansi* is one of few herbs approved as an acetylcholinesterase inhibitor from Ayurvedic medicines. Among this cohort of herbs, *jatamansi* was one of the most effective at inhibiting acetylcholinesterase activity *in vitro* (Mukherjee et al., 2007) thus making it a good candidate for an effective treatment for AD. When implicated *in vivo, jatamansi* could, in a chronic fatigue-induced model of Alzheimer's disease, improve multiple behavioral phenotypes including memory enhancement and cognitive parameters (Rahman et al., 2010).

Although evidence for the role of *jatamansi* in direct models of AD is lacking, the promising protective ability and mechanistic implications on the cholinergic systems makes *jatamansi* an attractive candidate for its treatment.

1.5.3 JATAMANSI IN PARKINSON'S DISEASE

Jatamansi has been previously shown to be a potent antioxidant and a strong inducer of biogenic amines. Pretreatment with an ethanolic extract of *jatamansi* significantly and dose-dependently rescued behavioral deficits induced by 6-OHDA, an oxidative model of PD. In addition, *jatamansi* extract dose-dependently prevented the development of a pro-oxidant state, the fall of dopamine levels in the substantia nigra and maintained the neuronal density of tyrosine-hydroxylase immunoreactive fibers (Ahmad et al., 2006).

Altogether, the efficacy of *jatamansi* in treating neurodegenerative diseases can be implied from its cholinergic stimulating activities and antioxidant potential, further studies are required to fully classify its effects. Though, based on its success and strong epidemiological success in Ayurveda, further studies are indeed warranted.

1.6 MUCUNA PRURIENS

Mucuna pruriens is known in Hindi as *kappichachu* or in English as cowhage. It has a variety of functions, though the most well known for its actions against PD. It is also a powerful reproductive rejuvenative supporting male potency and female fertility. Today, it is also used to treat anxiety, arthritis, infections, pain, fever and even snake bites.

Mucuna pruriens is an annual climbing shrub found in the Caribbean, Africa and India. It has long finger-like vines that can reach over 15 m. The leaves are large and ovate shaped, often highly grooved and pointy at the tips. There are also small flowers on this plant that appear purplish or white. The fruit of this plant that has a characteristic hairy look and carries several seeds. When this plant is young, it is almost completely covered in hair though this characteristic is lost in older plants. Be warned, if you touch the attractive hairy seedpod, you are liable to get an extreme itch as the hairs are enriched in serotonin.

Since the dawn of Ayurveda, *Mucuna pruriens* has been used for treating PD ailments, better known in Ayurveda as *kampavata*. Even today, the contemporary treatment for PD is an isolate from *Mucuna pruriens* seeds. This plant is categorized as a *Balya* or body-strengthening herb and is also a powerful *rasayan*. It has both sweet and bitter *rasa*, a hot *virya* and a sweet *vipaka*. This plant is best used to balance *vata* persons, can be aggravating to pitta is definitely aggravating *kapha* individuals. As PD is namely a disease of *vata vikriti*s, it is suitable that *Mucuna pruriens* mostly affects this *dosha*.

Mucuna pruriens is famous for being a natural source of L-Dopa (Modi et al., 2008), the precursor to dopamine that is commonly used in allopathic medicine as a treatment against PD. The *Mucuna pruriens* plant consists of about 4–6% L-Dopa, or 40 mg/g of the plant. It also contains several other neurotransmitters, namely serotonin, nicotine, DMT, bufotenine and 5-MeO-DMT. The abundance of such neurological effectors gives *Mucuna pruriens* a potent battery of psychedelic effects. *Mucuna pruriens* also contains a variety of alkaloids such as mucunin, mucunadine, mucuadinine, pruriendine, and nicotine.

Importantly, *Mucuna pruriens* has been shown to be non-toxic in animals and does not display any genotoxic effects on DNA (Manyam and Parikh, 2003; Dhanasekaran et al., 2008). Also interestingly, the bioavailability of L-Dopa from *Mucuna pruriens* seems to be higher in humans than normal L-Dopa treatment as there is a shorter latency time to peak L-Dopa concentrations in the plasma after oral administration (Katzenschlager et al., 2004). Based on the beneficial array of neurological modifying constituents of *Mucuna pruriens*, it is no doubt that this herb directly communicates with the central nervous system and has very influential effects there.

1.6.1 MUCUNA PRURIENS IN PARKINSON'S DISEASE

Mucuna pruriens is likely the most revered anti-PD treatment from Ayurveda known today. So nature has done it again, millennia before the creation of the synthetic levodopa drug, it had already conceived the perfect anti-PD medicine. Though, despite only have a small percentage of L-Dopa (4–6%), *Mucuna pruriens* is more efficient than L-Dopa treatment in alleviating the motor symptoms of PD (Hussain and Manyam, 1997). It also can potently restore the levels of dopamine in the substantia nigra in animals (Manyam et al., 2004) indicating the efficacy of the L-Dopa content. Notably, *Mucuna pruriens* has a unique property in that it does not instigate drug-induced dyskinesia, the side effect of L-Dopa treatment that inevitably limits its utility in treating PD patients (Lieu et al., 2010).

Mucuna pruriens, like any natural product, consists of many active ingredients. In addition to its anti-PD activity mediated by L-Dopa, many studies have attested to its anti-oxidative activity that also fights the symptoms of PD. Indeed *Mucuna pruriens* has a potent ability to scavenge reactive oxygen species and inhibit the oxidation of lipids (Dhanasekaran et al., 2008). In addition, it was shown *in vivo* that *Mucuna pruriens* inhibits liver lipid peroxidation and increases superoxide dismutase and catalase levels in rats (Agbafor et al., 2011).

It was also shown the *Mucuna pruriens* reduces many oxidative markers in a paraquat model of PD in mice. In particular, *Mucuna pruriens* reduced the levels of lipid peroxidation and increased the levels of catalase and superoxide dismutase. Behaviorally, *Mucuna pruriens* reduced the PD motor phenotypes and biochemically, there was an increase in dopaminergic staining in the substantia nigra demonstrating fundamental neuroprotective effects (Yadav et al., 2013).

In rats, chronic exposure to a water extract of *Mucuna pruriens* seed powder was shown to reduce the onset of dyskinesia in the 6-OHDA model of PD (Lieu et al., 2010). Interestingly, the reduction in dyskinesia was noted in animals given *Mucuna pruriens* without any additives, namely dopa-decarboxylase inhibitors. This suggests that there are some additional components in the water extract of *Mucuna pruriens* that alleviate the need for such additives that are normally necessary for the action of L-Dopa, or it is some other component besides L-Dopa that is mediating the anti-PD effects.

HP-200 is a commercially available formulation of *Mucuna pruriens*, which has been put through clinical trials in human PD patients. Indeed, It was shown to be effective in treating the symptoms of PD in these patients (HP-200 in Parkinson Disease Study Group, 1995). In addition, there have been several studies conducted in humans indicating that *Mucuna pruriens* seed powder extract (45 g/day) has significant improvements in PD symptoms after 12–20 weeks (HP-200, Vayda et al., 1978; Katzenschlager et al., 2004).

Thus, both clinically and biochemically, *Mucuna pruriens* has been shown to have beneficial effects in the fight against PD. It is interesting to note that the L-Dopa content may not be the only constituent that is offering the neurological benefits exemplifying why natural products having a broad range of active ingredients can be better medicines than isolated synthetic products.

1.7 SHATAVARI

Of the 300 species of asparagus, *Asparagus racemosus* (*shatavari*) has been demonstrated in multiple branches of alternative medicine to have potent medicinal properties. Do not be deceived, but the medicinally active 'asparagus' is different than that normally consumed as a popular vegetable in Europe and the Americas. The latter variety is characterized by its long green shoot and is accurately known as *Asparagus officinalis*. In Sanskrit, *shatavari* literally means "she who possesses a hundred husbands" as it was traditionally used as an aphrodisiac, especially for females.

Shatavari grows in tropical and subtropical parts of India up to an altitude of 1500 m. It is a small spiny under-shrub with a small root shaft bearing many small thick tuberous roots. These roots are lightly colored, white or ash-colored, with a white interior. The stem is woody with numerous small spines and it has mildly fragrant flowers. This plant also has small red fruits that darken upon maturity. Unfortunately, due to destruction, demand and deforestation, *shatavari* has become an endangered species naturally in the environment.

In Ayurveda, *shatavari* is best at pacifying disorders of *pitta* and has a *madhura* (sweet) and *tikta* (bitter) *rasa* with a cooling *virya*. *Shatavari* is a potent *rasayan* meaning that it promotes general wellness, longevity and strength (Dahanukar et al., 2000) and has powerful rejuvenative properties in several physiological systems. Typically, *shatavari* is used in women as a stimulator of fertility, regulator of menstruation and depressant of symptoms of menopause. It even prepares the body for pregnancy and protects against miscarriages. In the ancient Ayurvedic texts, *shatavari* was listed as having neuroprotective abilities therefore classified as a *medhya rasayan* (Warrier et al., 2000) and powerful nervine tonic (Parihar and Hemnani, 2004). Other actions of *shatavari* include *nidrajnana* (promoter of sleep) and *manasrogaghna* (alleviate mental disease) (Pandey et al., 1991; Sharma et al., 2001).

The medicinal properties of *shatavari* are mainly attributed to its roots. Like any natural product, *shatavari* roots contain a broad array of active ingredients, including steroidal saponins (shatavarins, saponins, immunoside), isoflavones, racemofuran, asparagamine A and racemosol (Hayes et al., 2006). Importantly, *shatavari* consists of a large proportion of phytoestrogens, which is the ingredient that makes it particularly effective at establishing female hormonal balance.

Today *shatavari* is known to have many functions including adaptogenic, digestive, antiulcerogenic, immunmodulatory, antibacterial, regulator of hormones, antioxidant, anti-depressive, antineoplastic and cardiovascular. It should be noted however, that many of these effects are concluded with formulations in which *shatavari* is just a part, so the entire effect cannot be solely attributed to *shatavari*. In fact, there have been very few studies investigating the isolated role of *shatavari*, though based on the success of its formulations, such tests are likely to arise in the near future.

One such popular polyherbal formulation is EuMil, which was designed according to Ayurvedic principles to be a powerful adaptogen. *Shatavari* is one of the main ingredients of EuMil but is also in combination with *Withania somnifera, Ocumum sanctum* and *Emblica officinalis* (Bhattacharya et al., 2002). Apart from allowing the body to react better to stressful situations, EuMil has potent effects on the central nervous system, namely the relief from depression and cognitive dysfunction (Muruganandam et al., 2002).

1.7.1 ANTIOXIDANT POTENTIAL OF SHATAVARI

Shatavari contains many antioxidant constituents such as ascorbic acid, polyphenols and flavonoids (Visavadiya and Narasimhacharya, 2005).

Indeed, the antioxidant potential of *shatavari* is comparable to glutathione and ascorbic acid as assessed by its ability to inhibit lipid peroxidation and protein oxidation (Kamat et al., 2000). *Shatavari* also directly induces the productivity of antioxidant enzymes such as superoxide dismutase and catalase (Bhatnagar et al., 2005; Visavadiya and Narasimhacharya, 2005).

1.7.2 NEUROPROTECTIVE EFFECTS OF SHATAVARI

Few studies have directly linked the actions of *shatavari* to neurodegeneration, but based on its anti-oxidantive actions in the central nervous system, *shatavari* has shown great potential to become a treatment for various neurodegenerative disorders. One study clearly outlined the neuroprotective effects of *shatavari* in both chronically stressed animals and humans with memory deficits and/or disturbances (Saxena et al., 2007). In another study, *shatavari* was shown to protect against neuronal death in the striatum and hippocampus in a kainic acid-induced excitotoxicity model. The ethanolic extract of *shatavari* attenuated oxidative damage by elevating glutathione peroxidase activity, glutathione content and by reducing lipid peroxidation (Parihar and Hemnani, 2004). Similar to the kainic acid induced insult, immobilization stress induced oxidative changes in the hippocampus, which were counteracted by the antioxidant effects of *shatavari* extracts (Vimal et al., 2010).

An important central nervous system-acting formulation containing *shatavari* is Mentat, which in addition to *shatavari* contains 20 other herbs; however the neurological effects are primarily attributed to the action of *Withania somnifera, Acorus calamus* and *shatavari*. Interestingly, in the 3-NP model of Huntington's disease, Mentat was able to effectively rescue behavioral deficits and the creatine kinase activity, which is characteristic of HD's pathology. In addition, the main individual components including *shatavari*, were shown to retain this effect in isolation (Patil et al., 2010).

To date, there are no specific studies dealing with the effect of *shatavari* on Parkinson's or Alzheimer's disease. However, *shatavari* has been long known to have potent anti-depressant activities indicated by *shatavari's* action on the monoaminergic systems, i.e. modulation of dopamine and serotonin pathways (Singh et al., 2009). Additionally, its nootropic and anti amnestic activities have been shown to be related to its acetyl cholinesterase inhibitory activity (Ojha et al., 2010) and its antioxidant activity (Gindi et al., 2011). These two observations of *shatavari* definitely link it to possible roles in protecting against Parkinson's and Alzheimer's diseases, respectively. Indeed, a methanolic extract of *shatavari* inhibited the acetylcholinesterase and dopamine catabolism (MAO-A and MAO-B) activity in cells (Meena et al., 2011) through a non-selective competitive inhibition model (Ojha et al., 2010). In addition, in a chronic stress model, EuMil normalized the perturbed monoamine neurotransmitter levels, namely dopamine and noradrenalin (Bhattacharya et al., 2002). Functionally, *shatavari* has been shown to protect against amnesia and certain memory deficits (Dhwaj et al., 2010). Thus, despite the lack of direct evidence linking *shatavari* to AD and PD, *in vitro* studies support its antioxidant potential and systematic implications in the pathological features of both these diseases.

1.8 SIDA CORDIFOLIA

Sida cordifolia ("*bala*" in Sanskrit or Country Mallow in English) belongs to the cotton family Malvaceae and is distributed throughout tropical and subtropical India. *Sida cordifolia* has been known for millennia in both Ayurvedic and Chinese medical systems and is commonly used for the treatment of stomatitis, asthma and nasal congestion (Balbach et al., 1978). However more recently, it has been shown in numerous studies to have potent antioxidant activity and to act on the central nervous system.

Sida cordifolia is a small plant, growing to about 1.5 meters in height with strong roots and stem with heart-shaped leaves and odorless yellow flowers. The roots are normally harvested for their medicinal activities though it has been shown that there are active ingredients in the leaves as well (Pole et al., 2006). Especially in the roots, this plant has been shown to have tonic properties, which help various nervous disorders such as hemiplegia and facial paralysis (Rastogi and Malhotra, 1985).

According to Ayurveda, *Sida cordifolia* is a general nutritive tonic (*rasayana*) acting to build tissues, strengthen the body and reduce general bodily stresses. In Ayurveda, *Sida cordifolia* has a *tikta* (bitter) and *madhura* (sweet) *rasa* with a *ruksha* (dry) and *guru* (heavy) *guna*, *madhura* (sweet) *vipaka* and a cooling *virya.* Due to the overly cooling nature of *Sida cordifolia,* it is best used in *pitta* individuals though is beneficial and balancing to all *doshas.* There are many varied traditional preparations of this herb, depending on the ailment to be treated and the *prakriti*. For example, oil preparations from a decoction of the root bark and mixed with milk and sesame oil is efficient in treating diseases of the nervous system (Koman et al., 1921). Alternatively, juice from the plant or a paste made from the roots can be applied topically for a variety of diseases (Jain et al., 2011).

As per its active ingredients, *S. cordifolia* contains mainly polyphenolic compounds, alkaloids, oils, steroids, resin acids, mucin and potassium nitrate (Pattar and Jayaraj, 2012; Diwan and Kanth, 1999). Notably, *Sida cordifolia* contains a significant proportion of ephedrine and psudoephedrine, stimulatory agents that suppress appetite, stimulate the mind and ultimately aid in weight loss (Jain et al., 2011).

The earliest pharmacological studies indicated that *Sida cordifolia* contained anti-inflammatory and analgesic activities (Antoniolli et al., 2000). In fact, the scope of *Sida cordifolia*'s action is broad. It has been reported to benefit cardiovascular health, lower blood pressure, control diabetes, supply antioxidants, aid in fat loss and even contribute to a cure for Parkinson's disease. It is also known as a cooling circulatory stimulant, which acts to increase blood flow without aggravating heat of *pitta dosha*. To support this, *Sida cordifolia* is also a demulcent, increasing the mucous secretions thus protecting the body from excess heat. Importantly, *Sida cordifolia* specifically tonifies the *mamsa dhatu* or the musculature of the body, which is a perfect compliment to aid in building after neurodegenerative assault. At very high doses, *Sida cordifolia* exhibits few toxic effects, and next to none if administered orally. If administered intraperitoneally, high doses can lead to sedation, reductions of spontaneous activity and a general depression phenotype (Quintans et al., 2005).

1.8.1 ANTIOXIDANT POTENTIAL OF SIDA CORDIFOLIA

Notably, *Sida cordifolia* is a cooling nervine tonic, acting to strengthen the nervous system and increase its stability. Indeed, it was one of the four original herbs used in a herbal concoction for the treatment of PD (Nagashayana et al., 2000). Of particular importance for battling neurodegenerative diseases is the justified antioxidant activity of *Sida cordifolia*. Initially, *in vitro* and *ex vivo* studies indicated that both ethanolic and water seed extracts from *Sida cordifolia* have a potent and dose-dependent antioxidant potential (Auddy et al., 2003)

1.8.2 *SIDA CORDIFOLIA* IN PARKINSON'S DISEASE

To investigate closely the ability of *Sida cordifolia* to attenuate the symptoms of PD, one group took various extracts (hexane, chloroform and aqueous) of *Sida cordifolia* and co-administered them to rats with rotenone treatment – a potent inducer of mitochondrial damage and oxidative stress known to elicit PD. Interestingly, *only* the aqueous extract was able to rescue postural and motor behaviors, reduce brain lesions and reduce oxidative damage through potentiating antioxidant systems. The aqueous extract even upregulated levels of dopamine in the brain whereas the other extracts failed to affect any of the parameters (Khurana et al., 2013).

It is clear that *Sida cordifolia* has great potential to combat all neurodegenerative diseases based on its innate properties. Further, with regards to the traditional uses of this plant in Ayurveda, further research regarding this effect is warranted.

KEYWORDS

- **Ayurveda**
- **herbs**
- **holistic**
- **neurodegenerative diseases**
- **oxidative**
- **rasayan**

REFERENCES

Abdel-Salam, O. M. Drugs used to treat Parkinson's disease, present status and future directions. CNS Neurol. Disord. Drug Targets. 2008, 7, 321–342.

Adams, B. K., Cai, J., Armstrong, J. EF24, a novel synthetic curcumin analog, induces apoptosis in cancer cells via a redoxdependent mechanism. Anticancer Drugs. 2005, 16, 263–275.

Agbafor, K. N., Nwachukwi, N. Phytochemical Analysis and Antioxidant Property of Leaf Extracts of Vitex doniana and Mucuna pruriens. Biochem. Res. Int. 2011, doi: 10.1155/2011/459839.

Ahlemeyer, B., Krieglstein, J. Neuroprotective effects of Ginkgo biloba extract. Cell Mol. Life Sci. 2003a, 60, 1779–1792.

Ahmad, M., Saleem, S., Ahmad, A. S., Ansari, M. A., Yousuf, S., Hoda, M.N. Neuroprotective effects of Withania somnifera on 6-hydroxydopamine induced Parkinsonism in rats. Hum. Exp. Toxicol. 2005, 24, 137–47.

Ahmad, M., Saleem, S., Ahmad, A.S., Yousuf, S., Ansari, M.A., Khan, M.B. Ginkgo biloba affords dose-dependent protection against 6-hydroxydopamine–induced parkinsonism in rats: neurobehavioural, neurochemical and immunohistochemical evidences. J. Neurochem. 2005, 93, 94–104.

Ahmad, M., Yousuf, S., Khan, M.B., Hoda, M. N., Ahmad, A.S., Ansari, M.A., Ishrat, T., Agrawal, A. K. Islam, E. Attenuation by Nardostachys jatamansi of 6 hydroxydopamine-induced parkinsonism in rats: behavioral, neurochemical and immunohistochemical studies. Pharmacol. Biochem. Behav. 2006, 83(1), 150–160.

Ahmed, M. E., Javed, H., Khan, M. M., Vaibhav, K., Ahmad, A., Khan, A., Tabassum, R., Islam, F., Safhi, M. M., Islam, F. Attenuation of oxidative damage-associated cognitive decline by Withania somnifera in rat model of strptozotocin-induced cognitive impairment. Protoplasma. 2013, 250, 1067–1078.

Alzheimer's Association: 2010 Alzheimer's disease facts and figures. Alzheimer's Dement. 2010, 6: 158–194.

Ananthacharya, E. Rasayana and Ayurveda. Ayurvediya Rasayana Kuti; Bezwada, 1939, p. 25.

Anbarasi, K., Vani, G., Balakrishna, K., Desai, C. S. Creatine kinase isoenzyme patterns upon chronic exposure to cigarette smoke: Protective effect of Bacoside A. Vascul. Pharmacol. 2005, 42, 57–61.

Antoniolli, A. R. Rodrigues, R. H. V. Mourao, M. R. Antiinflammatory, analgesic activity and acute toxicity of Sida cordifolia L. (Malva-branca). Journal of Ethnopharmacology. 2000, 72, 273–278.

Apparao, M. V. R., Srinivasan, K., Rao, K. Effect of Mandookaparni (Centella asiatica) on the general mental ability (Medhya) of mentally retarded children. J. Res. Ind. Med. 1973, 8, 9–16.

Arai, T., Fukae, J., Hatano, T. Up-regulation of hMUTYH, a DNA repair enzyme, in the mitochondria of substantia nigra in Parkinson's disease. Acta. Neuropathol. 2006, 112, 139–145.

Asahina, M., Shinotoh, H., Hirayama, K. Hypersensitivity of cortical muscarinic receptors in Parkinson's disease demonstrated by PET. Acta. Neurol. Scand. 1995, 91, 437–443.

Auddy, B., Ferreira, M., Blasina, F., Lafon, L., Arredondo, F., Dajas, F., Tripathi, P. C., Seal, T., Mukherjee, B. Screening of antioxidant activity of three Indian medicinal plants traditionally used for the management of neurodegenerative diseases. J. Ethical. Pharmacol. 2003, 84, 131–138.

Bagchi, A., Oshima, Y., Hikino, H. Neolignans and Lignans of Nardostachys jatamansi Roots. Planta Med. 1991, 57, 96–97.

Balbach, A. A Flora Medicinal and Medicina Domestica, vol. 2. 1978, MVP, Itaquaquecetuba, p. 703.

Bantubungi, K., Blum, D., 2007. Mechanisms of neuronal death in Huntington's disease. First part: general considerations and histopathological features. Revue medicale de Bruxelles. 28, 413–421.

Bartus, R. T., Dean, R. L. 3rd, Beer, B., Lippa, A. S. The cholinergic hypothesis of geriatric memory dysfunction. Science. 1982, 217(4558), 408–414.

Baum, L., Lam, C. W., Cheung, S. K. Six-month randomized, placebo-controlled, double-blind, pilot clinical trial of curcumin in patients with Alzheimer disease. J. Clin. Psychopharmacol. 2008, 28(1), 110–113.

Bhatnagar, M., Sharma, D., Salvi, M. Neuroprotective Effects of Withana somnifera Dunal: A Possible Mechanism. Neurochem. Res. 2009, 34, 1975–1983.

Bhatnagar, M., Sisodia, S. S., Bhatnagar, R. Antiulcer and antioxidant activity of Asparagus racemosus Willd and Withania somnifera Dunal in rats. Ann. N. Y. Acad. Sci. 2005, 1056, 261–278.

Bhattacharya, A., Ghosal, S., Bhattacharya, S. K. Anti-oxidant effect of Withania somnifera glycowithanolides in chronic foot shock stress induced perturbations of oxidative free radical scavenging enzymes and lipid peroxidation in rat frontal cortex and striatum. J. Ethnopharmacol. 2001, 74, 1–6.

Bhattacharya, A., Murugandam, A. V., Kumar, V., Bhattacharya, S. K. Effect of polyherbal formulation, EuMil, on neurochemical perturbations induced by chronic stress. Indian Journal of Experimental Biology. 2002, 40, 1161–1163.

Bhattacharya, A., Murugandam, A. V., Kumar, V., Bhattacharya, S. K. Effect of polyherbal formulation, EuMil, on neurochemical perturbations induced by chronic stress. Indian J. Exp. Biol. 2012, 40, 1161–1163.

Bhattacharya, A., Ramanathan, M., Ghosal, S. Effect of Withania somnifera glycowithanolites on iron induced hepatotoxicity in rats. Phytother Res 2000, 14, 568–570.

Bhattacharya, S. K., Bhattacharya, A., Kumar, A., Ghosal, S. Antioxidant activity of Bacopa monniera in rat frontal cortex, striatum and hippocampus. Phytother. Res. 2000, 14, 174–179.

Bhattacharya, S. K., Kumar, A., Ghosal, S. Effect of Bacopa monnieri on animal models of Alzheimer's disease and perturbed central cholinergic markers of cognition in rats. In: Siva Sankar DV (ed.) 1999 Molecular aspects of Asian medicines. PJD Publications, New York.

Bhattacharya, S. K., Kumar, A., Ghosal, S. Effects of glycowithanolides from Withania somnifera on animal model of Alzheimer's disease and perturbed central cholinergic markers of cognition in rats. Phytother Res. 1995, 9, 110–113.

Bhattacharya, S. K., Satyan, K. S., Ghosal, S. Antioxidant activity of glycowithanolides from Withania somnifera. Indian J. Exp. Biol. 1997, 35(3), 236–239.

Birks, J., Grimley, E. V., Van Dongen, M. Ginkgo biloba for cognitive impairment and dementia. Cochrane Database Syst. Rev. 2002, CD003120.

Bogdanov, M. B., Ferrante, R. J., Kuemmerle, S., Klivenyi, P., & Beal, M. F. Increased vulnerability to 3-nitropropionic acid in an animal model of Huntington's disease. J. Neurochem. 1998, 71(6), 2642–2644.

Braquet, P. Proofs of involvement of PAF-acether in various immune disorders using BN 52021 (ginkgolide B): a powerful PAF-acether antagonist isolated from Ginkgo biloba L. Adv. Prostaglandin Thromboxane Leukot. Res. 1986, 16, 179–198.

Brown, J. H., Taylor, P. Muscarinic receptors agonist and antagonists. In: Hardman, J. G., Limbird, L. E., Gilman's, A. G. (eds.) Goodman and Gilman's the pharmacological basis of therapeutics, 10th edn. McGraw-Hill, New York, p 10, 2010.

Butterfield, D. A., Boyd-Kimball, D. Amyloid beta-peptide (1–42) contributes to the oxidative stress and neurodegeneration found in Alzheimer disease brain. Brain Pathol. 2004, 14(4), 426–432.

Cai, Z., Zhao, B., Ratka, A. Oxidative stress and beta-amyloid protein in Alzheimer's disease. Neuromolecular Med. 2011, 13(4), 223–250.

Calabrese, C., Gregory, W. L., Leo, M., Kraemer, D., Bone, K., Oken, B. Effects of a standardized Bacopa monnieri extract on cognitive performance, anxiety, and depression in the elderly: a randomized, double-blind, placebo-controlled trial. J. Altern. Complement Med. 2008, 14, 707–713.

Cao, G., Russell, R. M., Lischner, N., Prior, R. L. Serum antioxidant capacity is increased by consumption of strawberries, spinach, red wine or vitamin C in elderly women. J. Nutr. 1998, 128, 2383–2390.

Cesarone, M. R., Laurora, G., De Sanctis, M. T., Belcaro, G. Activity of Centella asiatica in venous insufficiency. Minerva Cardioangiol. 1992, 40, 137–143.

Charak Samhita 6000 BC: Charaka translation into English: Translator: Shree Gulabkunverba Ayurvedic Society. Jamnagar, India: 1949.

Chatterji, A., Prakashi, S. C. New Delhi: National Institute of Science Communication. The Treatise on Indian Medicinal Plants. 1997, 5, 99–100.

Chaurasia, S. S., Panda, S., Kar, A. Withania somnifera root extract in the regulation of lead-induced oxidative damage in male mouse. Pharmacol. Res. 2000, 41(6), 663–666.

Checkoway, H., Powers, K., Smith-Weller, T., Franklin, G. M., Longstreth Jr., W. T., Swanson, P. D. Parkinson's disease risks associated with cigarette smoking, alcohol consumption, and caffeine intake. Am. J. Epidemiol. 2002, 155, 732–738.

Chopra, R. N., Nayar, S. L., Chopra, I. C. Glossary of Indian Medicinal Plants. Council of Scientific and Industrial Research: New Delhi 1956.

Choudhary, M. I., Nawaz, S. A., Ul-Haq, Z., Lodhi, M. A., Ghayur, M. N., Jalil, S. Withanolides a new class of natural cholinesterase inhibitors with calcium antagonistic properties. Biochem. Biophys. Res. Commun. 2005, 334, 276–287.

Chung, S., Sonntag, K. C., Anderson, T., Bjorklund, L. M., Park, J. J., Kim, D. M. Genetic engineering of mouse embryonic stem cells by Nurr1 enhances differentiation and maturation into dopaminergic neurons. Eur. J. Neurosci. 2002, 16, 1829–1838.

Colciaghi, F., Borroni, B., Zimmermann, M., Bellone, C., Longhi, A., Padovani, A., Cattabeni, F., Christen, Y., Di Luca, M. Amyloid precursor protein metabolism is regulated toward alpha-secretase pathway by Ginkgo biloba extracts. Neurobiol. Dis. 2004, 16, 454–460.

Commenges, D., Scotet, V., Renaud, S., Jacqmin-Gadda, H., Barberger-Gateau, P., Dartigues, J. F. Intake of flavonoids and risk of dementia. Eur. J. Epidemiol. 2000, 16, 357–363.

Dahanukar, S. A., Kulkarni, R. A., Rege, N. N. Pharmacology of medicinal plants and natural products. Indian J. Pharmacol. 2000, 32, S81–S118.

Danielson, S. R., Andersen, J. K. Oxidative and nitrative protein modifications in Parkinson's disease. Free Radic. Biol. Med. 2008, 44, 1787–94.

Dave, U. P., Chauvan, V., Dalvi, J. Evaluation of BR-16 A (Mentat) in cognitive and behavioral dysfunction of mentally retarded children: A placebo-controlled study. Indian J. Pediatr. 1993, 60, 423–428.

de Rijk, M. C., Breteler, M. M., den Breeijen, J. H., Launer, L. J., Grobbee, D. E., van der Meche, F. G., Hofman, A. Dietary antioxidants and Parkinson disease. The Rotterdam Study. Arch. Neurol. 1997, 54, 762–765.

Deepak, M., Amit, A. The need for establishing identities of bacoside A and B, the putative major bioactive saponins of Indian medicinal plant Bacopa monnieri. Phytomedicine 2004, 11, 264–8.

DeFeudis, F. V., Drieu, K. Ginkgo biloba extract (EGb 761) and CNS functions: basic studies and clinical applications. Curr. Drug Targets. 2000, 1, 25–28.

Dexter, D. T., Carter, C. J., Wells, F. R., Javoy-Agid, F., Agid, Y., Leesz, A. Basal lipid peroxidation in substantia nigra is increased in Parkinson's disease. J. Neurochem. 1989, 52, 381–389.

Dhanasekaran, M., Tharakan, B., Holcomb, L. A., Hitt, A. R., Young, K. A., Manyam, B. V. Neuroprotective mechanisms of Ayurvedic antidementia botanical Bacopa monnieri. Phytother. Res. 2007, 21.

Dhanasekaran, M., Tharakan, B., Manyam, B. V. Antiparkinson Drug – Mucuna pruriens shows Antioxidant and Metal Chelating Activity. Phytother. Res. 2008, 22, 6–11.

Dhingra, D., Goyal, P. K. Inhibition of MAO and GABA: Probable mechanisms for antidepressant-like activity of Nardostachys jatamansi DC in mice. Indian J. Exp. Biol. 2008, 46, 212–218.

Dhwaj, A. V., Singh, R. Reversal effect of Asparagus Racemosus Wild (Liliaceae) Root extract on Memory deficits of Mice. Int. J. Drug Dev. Res. 2011, 3(2), 314–323.

Dickon, T. C., Vickers, J. C. The morphological phenotype of b-amyloid plaques and associated neuritic changes in Alzheimer's disease. Neuroscience. 2001, 105:99–107.

Diwan, P. V., Kanth, V. R. Analgesic, antiinflammatory and hypoglycaemic activities of Sida cordifolia. Phytotherapy Res. 1999, 13, 75–77.

Drieu K. Preparation and definition of G. biloba extract. Press. Med. 1986, 15, 1455–1457.

Dugaheh, M. A., Meisami, F., Torabian, Z., Sharififar, F. Antioxidant effect and study of bioactive components of Valeriana sisymbriifolia and Nardostachys jatamansii in comparison to Valeriana offcinalis. Pak. J. Pharm. Sci. 2013, 26(1), 53–58.

El Jaber-Vazdekis, N., Gonzalez, C., Ravelo, A. G. Cloning, characterization and analysis of expression profiles of a cDNA encoding a hyoscyamine 6-beta-hydroxylase (H6H) from Atropa baetica Willk. Plant Physiol. Biochem. 2009, 47, 20–25.

Ganguli, M., Chandra, V., Kamboh, M. I. Apolipoprotein E polymorphism and Alzheimer disease: The Indo-US Cross-National Dementia Study. Arch. Neurol. 2000, 57, 824–830.

Gil, J. M., Rego, A. C. Mechanisms of neurodegeneration in Huntington's disease. Eur. J. Neurosci. 2008, 27, 2803–2820.

Gilani, A. H., Khan, A. U., Raoof, M. Gastrointestinal, selective airways and urinary bladder relaxant effects of Hyoscyamus reticulatus are mediated through dual blockade of muscarinic receptors and Ca2 channels. Fundam. Clin. Pharmacol. 2008, 22, 87–99.

Gindi, S. Evaluation of nootropic potential and *in vitro* antioxidant activity of aqueous extract of roots of Aparagus racemosus in rats. Int. J. Pharm. Res. Dev. 2011, 3, 184–191.

Goehler, H., Lalowski, M., Stelzl, U. A protein interaction network links GIT1, an enhancer of Huntingtin aggregation, to Huntington's disease. Mol. Cell. 2004, 15(6), 853–865.

Gohil, K., Packer, L. Global gene expression analysis identifies cell and tissue specific actions of Ginkgo biloba extract, EGb 761. Cell Mol. Biol. 2002, 48, 625–631.

Gohil, K. J., Patel, J. A., Gajjar, A. K. Pharmacological Review of Centella asiatica: A Potential Herbal Cure-all. Indian J. Pharmac. Sci. 2010, Sept-Oct, 546–557.

Gould, T. J., Bowenkamp, K. E., Larson, G., Zahniser, N. R., Bickford, P. C. Effects of dietary restriction on motor learning and cerebellar noradrenergic dysfunction in aged F344 rats. Brain Research. 1995, 684(2), 150–158.

Grin, I. R., Konorovsky, P. G., Nevinsky, G. A., Zharkov, D. O. Heavy metal ions affect the activity of DNA glycosylases of the fpg family. Biochemistry (Mosc). 2009, 74: 1253–1259.

Guo, J. S., Cheng, C. L., Koo, M. W. Inhibitory effects of Centella asiatica water extract and asiaticoside on inducible nitric oxide synthase during gastric ulcer healing in rats. Planta. Med. 2004, 70, 1150–1154.

Gupta, R., Flora, S. J. Effect of Centella asiatica on arsenic induced oxidative stress and mental distribution in rats. J. Appl. Toxicol. 2006, 26, 213–222.

Gupta, S. K., Dua, A., Vohra, B. P. Withania somnifera (Ashwagandha) attenuates anti-oxidant defense in aged spinal cord and inhibits copper induced lipid peroxidation and protein oxidative modifications. Drug Metabol. Drug Interact. 2003, 19, 211–222.

Hagemann, R. C., Burnham, T. H., Granick, B., Neubauer, D. Gotu Kola, In, The Lawrence Review of Natural Products: facts and comparisons. St. Louis, MO, Facts and Comparisons Division, J. B. Lippincott Co., 1996. p. 41–2.

Haleagrahara, N., Ponnusamy, K. Neuroprotective effect of Centella asiatica extract (CAE) on experimentally induced parkinsonism in aged Sprague-Dawley rats. J. Toxicol. Sci. 2010, 35(1), 41–47.

Halliwell, B. Free radicals and antioxidants – quo vadis? Trends Pharmacol. Sci. 2011, 32, 125–30.

Hardy, J. A., Higgins, G. A. Alzheimer's disease: the amyloid cascade hypothesis. Science 1992, 256, 184–185.

Hayes, P. Y., Jahidin, A. H., Lehmann, R. P., Penman, K, Kitching, W., De Voss, J. J. Structural revision of Shatavarin I and IV, the major components from the roots of A. racemosus, Tetrahedron Lett. 2006, 47, 6965–6969.

Hickey, M. A., Zhu, C., Medvedeva, V. Improvement of neuropathology and transcriptional deficits in CAG 140 knock-in mice supports a beneficial effect of dietary curcumin in Huntington's disease. Mol. Neurodegen. 2012, 7, 12.

Hosamani, R., Muralidhara. Neuroprotective efficacy of Bacopa monnieri against rotenone induced oxidative stress and neurotoxicity in Drosophila melanogaster. Neurotoxicology. 2009, 30, 977–985.

Hosamani, R., Muralidhara. Prophylactic treatment with Bacopa monnieri leaf powder mitigates paraquat-induced oxidative perturbations and lethality in Drosophila melanogaster. Indian J. Biochem. Biophys. 2010, 47, 75–82.

HP-200 in Parkinson's Disease Study Group. An alternative medicine treatment for Parkinson's disease: results of a multicenter clinical trial. J. Altern. Complement. Med. 1995, 1, 249–255.

Huang, H. C., Chang, P., Dai, X. L., Jiang, Z. F. Protective effects of curcumin on amyloid-β-induced neuronal oxidative damage. Neurochem. Res. 2012, 37(7), 1584–1597.

Hussain, G., Manyam, B. V. Mucuna pruriens proves more effective than L-DOPA in Parkinson's disease animal model. Phytother. Res. 1997, 11, 419–423.

Inamdar, P. K., Yeole, R. D., Ghogare, A. B., de Souza, N. J. Determination of biologically active constituents in Centella asiatica. J. Chromatogr. A. 1996, 742, 127–130.

Iwunze, M. O., McEwan, D. Peroxynitrite interaction with curcumin solubilized in ethanolic solution. Cell Mol. Biol. (Noisy-le-grand). 2004, 50, 749–752.

Jadiya, P., Khan, A., Sammi, S. R., Kaur, S., Mir. S.S., Nazir, A. Anti-Parkinsonian effects of Bacopa monnieri: Insights from transgenic and pharmacological Caenorhabditis elegans models of Parkinson's disease. Biochem. Biophys. Res. Comm. 2011, 413, 605–610.

Jain, A. Choubev, S. Singour, P. K. Rajak, H. Pawar, R. S. Sida Cordifolia (Linn) – An Overview. J. Appl. Pharmaceu. Sci. 2011, 01(02), 23–31.

JanBen, I., Sturtz, S., Skipka, G., Zentner, A., Garrido, M., Busse, R. Ginkgo bloba in Alzheimer's disease: a systemic review. Wien. Med. Wochenschr. 2010, 21–22, 539–546.

Jayashree, G., Kurup, M.G., Sudarslal, V. S., Jacob, V. B. Anti-oxidant activity of Centella asiatica on lymphoma-bearing mice. Fitoterapia. 2003, 74, 431–434.

Joshi, H., Parle, M. Nardostachys jatamansi improves learning and memory in mice. J. Med. Food Spring. 2006, 9(1), 113–118.

Joshi, H., Parle, M. Nardostachys jatamansi improves learning and memory in mice. J. Med. Food. 2006, 9, 113–188.

Kamat, J. P., Boloor, K. K., Devasagayam, T. P. A., Venkatachalam, S. R. Antioxidant properties of Asparagus racemosus against damage induced by gamma-radiation in rat liver mitochondria. J. Ethnopharmacol. 2000, 71, 425–435.

Kang, X., Chen, J., Xu, Z., Li, H., Wang, B. Protective effects of Ginkgo biloba extract on paraquat-induced apoptosis of PC12 cells. Toxicol. *In Vitro*. 2007, 21, 1003–1009.

Karkada, G., Shenoy, K. B., Karanth, K. S. Nardostachys jatamansi extract prevents chronic restraint stress-induced learning and memory deficits in the radial arm maze task. J. Nat. Sci. Biol. Med. 2012, 3(2), 125–132.

Katzenschlager, R., Evans, A., Manson, A., Patsalos, P. N., Ratnaraj, N., Watt, H., Timmermann, L., Van der Giessen, R., Lees, A. J. Mucuna pruriens in Parkinson's disease: a double blind clinical and pharmacological study. J. Neurol. Neurosurg. Psychiatry. 2004, 75, 1672–1677.

Khurana, N., Gajbhiye, A. Ameliorative effect of Sida cordifolia in rotenone induced oxidative stress model of Parkinson's disease. Neurotox. 2013, 39, 57–64.

Kim, M. S., Lee, J. I., Lee, W. Y., Kim, S. E. Neuroprotective effect of Ginkgo biloba L. extract in a rat model of Parkinson's disease. Phytother. Res. 2004, 18, 663–666.

Kolodziejczyk, J., Olas, B., Saluk-Juszczak, J., Wachowicz, B. Antioxidative properties of curcumin in the protection of blood platelets against oxidative stress *in vitro*. Platelets. 2011, 22(4), 270–276.

Koman. Indian Materia Medica Gazette. August 1921.

Konar, A., Shah, N., Singh, R., Saxena, N., Kaul, S., Wadhwa, R., Thakur, M. Protective Role of Ashwagandha Leaf Extract and Its Component Withanone on Scopolamine-Induced Changes in the Brain and Brain-Derived Cells. PLoS One. 2011, 6(11), e27265.

Kuboyam, A. T., Tohda, C., Komatsu, K. Withanoside IV and its active metabolite, sominone, attenuate A beta(25–35)-induced neurodegeneration. Eur. J. Neurosci. 2006, 23(6), 1417–1426.

Kuboyama, T., Tohda, C., Komatsu, K. Neuritic regeneration and synaptic reconstruction induced by withanolide A. Br. J. Pharmacol. 2005, 144, 961–97.

Kuboyama, T., Tohda, C., Zhao, J., Nakamura, N., Hattori, M., Komatsu, K. Axon- or dendrite-predominant outgrowth induced by constituents from Ashwagandha. Neuroreport. 2002, 13, 1715–1720.

Kulkarni, S. K., Sharma, A., Verma, A., Ticku, M. K. GABA receptor mediated anticonvulsant action of Withania somnifera root extract. Indian Drugs. 1993, 30, 305–312.

Kumar, A., Naidu, P. S., Seghal, N., Padi, S. S. Effect of curcumin on intracerebroventricular colchicine-induced cognitive impairment and oxidative stress in rats. J. Med. Food. 2007, 10, 486–494.

Kumar, A., Prakash, A., Dogra, S. Centella asiatica attenuated D-Galactose Induced Cognitive Impairment, Oxidative and Mitochondrial Dysfunction. Intl. J. Alz. Dis. 2011, doi: 10.4061/2011/347569.

Kumar, P., Kumar, A. Effects of root extract of Withania somnifera in 3 nitropropionic acid-induced cognitive dysfunction and oxidative damage in rats. Int. J. Health Res. 2008, 1, 139.

Kumar, P., Kumar, A. Possible neuroprotective effect of Withania somnifera root extract against 3-nitropropionic acid-induced behavioral, biochemical and mitochondrial dysfunction in an animal model of Huntington's disease. J. Med. Food. 2009, 12(3), 591–600.

Kumar, V. M. H., Gupta, Y. K. Effect of different extracts of Centella asiatica on cognition and markers of oxidative stress in rats. J. Ethnopharmacol. 2002, 79, 253–260.

Kuppurajan, K., Srinivasan, K., Janaki, K. A double blind study of the effect of Mandukaparni on the general mental ability of normal children. J. Res. Indian Med. Yoga Homoeo. 1978, 13, 37–41.

Lahiri, D. K., Rogers, J. T., Greig, N. H., Sambamurti, K. Rationale for the development of cholinesterase inhibitors as anti-Alzheimer agents. Curr. Pharmac. Design. 2004, 10, 3111–3119.

Lautenschlager, N. T., Ihl, R., Muller, W. E. Ginkgo biloba extract EGb 761 in the context of current developments in the diagnosis and treatment of age-related cognitive decline and Alzheimer's disease: a research perspective. Intl. Psychogeriatrics. 2012, 24, S46–550.

Lees, A. J., Hardy, J., Revesz, T. Parkinson's disease. Lancet. 2009, 373(9680): 2055–2066.

Lieu, C. A., Kunselman, A. R., Manyam, B. V., Venkiteswaran, K., Subramanian, T. A water extract of Mucuna pruriens provides long-term amelioration of parkinsonism with reduced risk for dyskinesias. Parkinsonism and Related Disorders. 2010, 16, 458–465.

Lim, G. P., Chu, T., Yang, F., Beech, W., Frautschy, S. A., Cole, G. M. The curry spice curcumin reduces oxidative damage and amyloid pathology in an Alzheimer transgenic mouse. J. Neurosci. 2001, 21(21), 8370–8377.

Limpeanchob, N., Jaipan, S., Rattanakaruna, S., Phrompittayarat, W., Ingkaninan, K. Neuroprotective effect of Bacopa monnieri on beta-amyloid-induced cell death in primary cortical culture. J. Ethnopharmacol. 2008, 120, 112–117.

Liu, M. R., Han, T., Chen, Y., Qin, L. P., Zheng, H. C., Rui, Y. C. Effect of madecassoside on depression behavior of mice and activities of MAO in different brain regions of rats. J. Chinese Integr. Med. 2004, 2, 440–444.

Liu, T. J., Yeh, Y. C., Ting, C. T., Le, W. L., Wang, L. C., Lee, H. W. Ginkgo biloba extract 761 reduced doxorubicin-induced apoptotic damage in rat hearts and neonatal cardiomyocytes. Cardiovasc. Res. 2008, 80, 227–235.

Lyle, N., Battacharyya, D., Sur, T. K., Munshi, S., Paul, S., Chatterjee, S., Gomes, A. Stress modulating antioxidant effect Nardostchys jatamansi. J. Biochem. Biophys. 2009, 46(1), 93–98.

Mahdy, H. M., Tadros, M. G., Mohamed, M. R., Karin, A. N., Khalifa, A. E. The effect of Ginkgo biloba extract on 3-nitropropionic acid-induced neurotoxicity in rats. Neurochem. Intl. 2011, 59, 770–778.

Maitra, I., Marcocci, L., Droy-Lefaix, M. T., Packer, L. Peroxyl radical scavenging activity of Ginkgo biloba extract EGb 761. Biochem. Pharmacol. 1995, 49, 1649–55.

Mandel, S., Grunblatt, E., Riederer, P. Neuroprotective strategies in Parkinson's disease: an update on progress. CNS Drugs 2003, 17, 729–762.

Manjunath, M. J., Muralidhara. Effect of Withania somnifera Supplementation on Rotenone-Induced Oxidative Damage in Cerebellum and Striatum of the Male Mice Brain. Cent. Nerv. Syst. Agents Med. Chem. 2013, 13, 43–56.

Manyam, B. V., Parikh, K. M. Antiparkinsonian activity of Mucuna pruriens seeds. Ann. Neurosci. 2003, 9, 40–46.

Manyam, B. V., Dhanasekaran, M., Hare, T. Neuroprotective effects of antiparkinson drug Mucuna pruriens. Phytother. Res. 2004, 18, 706–712.

Marcocci, L., Packer, L., Droy-Lefaix, M., Sekaki, A., Gardes-Albert, M. Antioxidant action of Ginkgo biloba extract EGb 761. Methods Enzymol. 1994, 234, 462–475.

Maxwell, S., Cruickshank, A., Thorpe, G. Red wine and antioxidant activity in serum. Lancet 1994, 344, 193–194.

Meena, J., Ojha, R., Muruganandam, A. V., Krishnamurthy, S. Asparagus racemosus competitively inhibits *in vitro* the acetylcholine and monoamine metabolizing enzymes. Neurosci. Lett. 2011, 503, 6–9.

Menon, V. P., Sudheer, A. R. Antioxidant and anti-inflammatory properties of curcumin. Adv. Exp. Med. Biol. 2007, 595, 105–125.

Mishra, L. C., Singh, B. B., Dagenais, S. Scientific basis for the therapeutic use of Withania somnifera. (Ashwagandha): A review. Altern. Med. Rev. 2000, 5, 334–346.

Mishra, S., Srivastava, S., Dwivedi, S., Tripathi, R. D. Investigation of biochemical responses of Bacopa monnieri L. upon exposure to arsenate. Environ. Toxicol. 2011, doi: 10.1002/tox.20733

Modi, K. P., Patel, N. M., Goyal, R. K. Estimation of L-dopa from Mucuna pruriens LINN and formulations containing M. pruriens by HPTLC method. Chem. Pharm. Bull. (Tokyo) 2008, 56, 357e9.

Mohandas Rao, K. G., Muddanna, R. S., Gurumadhava, R. S. Centella asiatica (L.) leaf extract treatment during the growth spurt period enhances hippocampal CA3 neuronal dendritic arborization in rats. Evid. Based Complement. Alternat. Med. 2006, 3, 349–357.

Mook-Jung, I., Shin, J. E., Yun, S. H., Huh, K., Koh, J. Y., Park, H. K., Protective effects of asiaticoside derivatives against beta-amyloid neurotoxicity. J. Neurosci. Res. 1999, 58, 417–425.

Mukherjee, P. K., Kumar, V., Houghton, P. J. Screening of Indian medicinal plants for acetylcholinesterase inhibitory activity. Phytother. Res. 2007, 21(12), 1142–1145.

Munduvelil, T. T. Rajani, K., Anil, J. J., Sreeja, P. C., Paravanparampil, J. M., Mathew, D., Sabulal, B. Elite genotypes/chemotypes with high contents of madecassoside and asiaticoside from sixty accessions of Centella asiatica of south India and the Andaman Islands:

For cultivation and utility in cosmetic and herbal drug applications. Industrial Crops and Products. 2010, 32, 545–550.
Murthy, P. B., Raju, V. R., Ramakrisana, T., Chakravarthy, M. S., Kumar, K. V., Kannababu, S. Estimation of twelve bacopa saponins in Bacopa monnieri extracts and formulations by high-performance liquid chromatography. Chem. Pharm. Bull. (Tokyo). 2006, 54, 907–11.
Mythri, R. B., Bharath, M. M. S. Curcumin: A Potential Neuroprotective Agent in Parkinson's Disease. Current Pharmaceut. Design. 2012, 18, 91–99.
Mythri, R. B., Harish, G., Dubey, S. K., Misra, K., Bharath, M. M. Glutamoyl diester of the dietary polyphenol curcumin offers improved protection against peroxynitrite-mediated nitrosative
Mythri, R. B., Jagatha, B., Pradhan, N., Andersen, J., Bharath, M. M. Mitochondrial complex I inhibition in Parkinson's disease: how can curcumin protect mitochondria? Antioxid Redox. Signal. 2007, 9, 399–408.
Nagashayana, N., Sankarankutty, P., Nampoothiri, M. R. V., Mohan, P. K., Mohanakumar, K. P. Association of L-DOPA with recovery following Ayurveda medication in Parkinson's disease. J. Neurol. Sci. 2000, 176, 124–127.
Nevado, J., Sanz, R., Sanchez-Rodrıguez, C., Garcia-Berrocal, J. R., Martın-Sanz, E., Gonzalez-Garcıa, J. A. Ginkgo biloba extract (EGb 761) protects against aging-related caspase-mediated apoptosis in rat cochlea. Acta. Otolaryngol. 2010, 130, 1101–1112.
Ojha, R., Sahu, A. N., Muruganandam, A. V., Singh, G. K., Sairam, K., Krishnamurthy, S. Asparagus recemosus enhances memory and protects against amnesia in rodent models. Brain Cogn. 2010, 74, 1–9.
Oken, B. S., Storzbach, D. M., Kaye, J. A. The efficacy of Ginkgo biloba on cognitive function in Alzheimer disease. Arch. Neurol. 1998, 55, 1409–1415.
Pandey, N., Strider, J., Nolan, W. C., Yan, S. X., Galvin, J. E. Curcumin inhibits aggregation of alpha-synuclein. Acta. Neuropathol. 2008, 115, 479–489.
Pandey, V. N. Medico-ethno botanical exploration in Sikkim Himalaya, Central Council for research in Ayurveda & Siddha, First edition, 1991, pg. 137–189.
Pardon, M. C., Joubert, C., Perez-Diaz, F., Christen, Y., Launay, J. M., Cohen- Salmon, C. In vivo regulation of cerebral monoamine oxidase activity in senescent controls and chronically stressed mice by long-term treatment with Ginkgo biloba extract (EGb 761). Mech. Ageing Dev. 2000, 113, 157–168.
Parihar, M. S., Hemnani, T. Experimental excitotoxicity provokes oxidative damage in mice brain and attenuation by extract of Asparagus racemosus. J. Neural Transm. 2004, 111, 1–12.
Patil, P. T., Majagi, S. I. Effect of Mentat and its selected ingredients on 3-Nitropropionic acid induced neuronal damage in wistar rats. Pharmacologyonline. 2010, 1, 856–863.
Pattar, P. V. Jayaraj, M. Pharmacognostic and phytochemical investigation of Sida cordifolial. – A threatened medicinal herb. Int. J. Pharm. Pharm. Sci. 2012, 4(1), 114–7.
Pole, Sebastian. Ayurvedic medicine. Elsevier Health Sciences. 2006, 137.
Potter, P. Curcumin: a natural substance with potential efficacy in Alzheimer's disease. J. Exp. Pharma. 2013, 5, 23–31.
Prakash, D., Suri, S., Upadhyay, G. Total phenol, antioxidant and free radical scavenging activities of some medicinal plants. Int. J. Food. Sci. Nutr. 2007, 58(1), 18–28.

Prakash, J., Yadav, S., Chouhan, S., Singh, S. Neuroprotective Role of Withania somnifera Root Extract in Maneb-Paraquat Induced Mouse Model of Parkinsonism. Neurochem. Res. 2013, 38, 972–980.

Praticò, D., Trojanowski, J. Q. Inflammatory hypotheses: novel mechanisms of Alzheimer's neurodegeneration and new therapeutic targets? Neurobiol. Aging. 2000, 21, 441–445.

Quik, M. Smoking, nicotine and Parkinson's disease. Trends Neurosci. 2004, 27, 561–568.

Quintans-Junior, L. J. Almeida, R. N. Antoniolli, A. R. CNS pharmacological effects of the hydroalcoholic extract of Sida cordifolia L. leaves. J. Ethical. Pharmacol. 2005, 98, 275–279

Rahman, H., Murlidharan, P. Nardostacys jatamansi DC Protects from the loss of Memory and Cognition Deficits in Sleep Deprived Alzheimer's Disease (AD) Mice Model. Intl. J. Pharma. Sci. Rev. Res. 2010, 5(3), 160–167.

RajaSankar, S., Manivasagam, T., Sankar, V., Prakash, S., Muthusamy, R., Krishnamurti, A., Surendran, S. Withania somnifera root extract improves catecholamine and physiological abnormalities seen in a Parkinson's disease model mouse. J. Ethnopharmacology. 2009, 125, 369–373.

Rajeswari, A., Sabesan, M. Inhibition of monoamine oxidase-B by the polyphenolic compound, curcumin and its metabolite tetrahydrocurcumin, in a model of Parkinson's disease induced by MPTP neurodegeneration in mice. Inflammopharmacology. 2008, 16, 96–99.

Rajeswari, A. Curcumin protects mouse brain from oxidative stress caused by 1-methyl-4-phenyl-1,2,3,6-tetrahydropyridine. Eur. Rev. Med. Pharmacol. Sci. 2006, 10, 157–161.

Ramasamy, C. Emerging role of polyphenolic compounds in the treatment of neurodegenerative diseases: A review of their intracellular targets. Eur. J. Pharmacol. 2006, 545, 51–64.

Ramassamy, C., Poirier, J. Ginkgo biloba extract (EGb 761) and apolipoprotein E isoforms on the ß-amyloid fibril formation. In: Christen, Y. (Ed.), Ginkgo Biloba Extract (EGb 761) and Neurodegenerative Diseases. 2001 pp. 71–91.

Ramassamy, C. Emerging role of polyphenolic compounds in the treatment of neurodegenerative diseases: A review of their intracellular targets. Eur. J. Pharmacol. 2006, 545, 51–64.

Ramassamy, C., Averill, D., Beffert, U., Bastianetto, S., Theroux, L., Lussier-Cacan, S., Cohn, J. S., Christen, Y., Davignon, J., Quirion, R., Poirier, J. Oxidative damage and protection by antioxidants in the frontal cortex of Alzheimer's disease is related to the apolipoprotein E genotype. Free Radic. Biol. Med. 1999, 27, 544–553.

Ramoutsaki, I. A., Papadakis, C. E., Ramoutsakis, I. A. Therapeutic methods used for otolaryngological problems during the Byzantine period. Ann. Otol. Rhinol. Laryngol. 2002, 111, 553–557.

Rao, S. B., Chetana, M., Devi, U. P. Centella asiatica treatment during postnatal period enhances learning and memory in mice. Physiol. Behav. 2005, 86, 449–457.

Rao, V. S., Rao, A., Karanth, K. S. Anticonvulsant and neurotoxicity profile of Nardostachys jatamansi in rats. J. Ethnopharmacol. 2005, 102, 351–356.

Rasheed, A. S., Venkataraman, S., Jayaveera, K. N., Fazil, A. M., Yasodha, K. J., Aleem, M. A., Mohammed, M., Khaja, Z., Ushasri, B., Pradeep, H. A., Ibrahim, M. Evaluation of toxicological and antioxidant potential of Nardostachys jayamansi in reversing haloperidol-induced catalepsy in rats. J. Gen. Med. 2010, 26(3), 127–136.

Rastogi, M., Ojha, R., Prabu, P., Parimala, S. B., Agrawal, A., Dubey, G. Prevention of age-associated neurodegeneration and promotion of healthy brain ageing in female Wistar rats by long-term use of bacosides. Biogerontology 2012, 13, 183–195.

Rastogi, R. P. Malhotra, B. N. Compendium of Indian medical plants. 1985, p. 674.

Ringman, J. M., Cole, G. M., Teng, E. Oral curcumin for the treatment of mild-to-moderate Alzheimer's disease: tolerability and clinical and biomarker efficacy results of a placebo-controlled 24-week study. Alzheimer's Dement. 2008, 4(Suppl 4), T774.

Rodriguez-Puertas, R., Pazos, A., Pascual, J. Cholinergic markers in degenerative parkinsonism: autoradiographic demonstration of high-affinity choline uptake carrier hyperactivity. 1994, Brain Res. 636, 327–332.

Rojas, P., Garduno, B., Rojas, C., Vigueras, R. M., Rojas-Castaneda, J., Rios, C., Serrano-Garcia, N. EGb 761 blocks MPPfl-induced lipid peroxidation in mouse corpus striatum. Neurochem. Res. 2001, 26, 1245–1251.

Rojas, P., Ruiz-Sanchez, E., Rojas, C., Ogren, S. O. Ginkgo biloba extract (EGb 761) modulates the expression of dopamine-related genes in the MPTP-Induced Parkinsonism in mice. Neurosci. 2012, 223, 246–257.

Rojas, P., Serrano-Garcıa, N., Mares-Samano, J. J., Medina-Campos, O. N., Pedraza-Chaverri, J., Ogren, S. O. EGb 761 protects against nigrostriatal dopaminergic neurotoxicity in 1-methyl-4-phenyl-1,2,3,6-tetrahydropyridine-induced Parkinsonism in mice: role of oxidative stress. Eur. J. Neurosci. 2008, 28, 41–50.

Roodenrys, S., Booth, D., Bulzomi, S., Phipps, A., Micallef, C., Smoker, J. Chronic effects of Brahmi (Bacopa monnieri) on human memory. Neuropsychopharmacology. 2002, 27, 279–281.

Rueker, G., Panicker, M. R., Breitamaier, E. Revised structure and stereochemistry of jatamansic oil. Phytochemistry. 1993, 33, 141–143.

Russo, A., Borrelli, F. Bacopa monnieri, a reputed nootropic plant: An overview. Phytomedicine. 2005, 12, 305–17.

Russo, A., Borrelli, F., Campisi, A., Acquaviva, R., Raciti, G., Vanella, A. Nitric oxide-related toxicity in cultured astrocytes: effect of Bacopa monnieri. Life Sci. 2003a, 73, 1517–1526.

Russo, A., Izzo, A. A., Borrelli, F., Renis, M., Vanella, A. Free radical scavenging capacity and protective effect on DNA damage of Bacopa monniera L. Phytother Res. 2003b, 17, 870–875

Saini, N., Singh, D., Sandhir, R. Neuroprotective effects of Bacopa monnieri in experimental model of dementia. 2012, 37(9), 1928–1937.

Sairam, K., Dorababu, M., Goel, R. K., Bhattacharya, S. K. Antidepressant activity of standardized extract of Bacopa monniera in experimental models of depression in rats. Phytomedicine. 2002, 9, 207–211.

Salim, S., Ahmad, M., Khan, S. Z., Ahmad, S. A., Islam, F. Protective effect of Nardostachys jatamansi in rat cerebral ischemia. Pharmacol. Biochem. Behav. 2003, 74, 481–486.

Sandhir, R., Yadav, A., Mehrotra, A., Sunkaria, A., Singh, A., Sharma, S. Curcumin Nanoparticles Attenuate Neurochemical and Neurobehavioral Deficits in Experimental Model of Huntington's Disease. Neurol. Med. 2013, doi: 10.1007/s12017–013–8261-y.

Sandur, S. K., Ichikawa, H., Pandey, M. K. Role of pro-oxidants and antioxidants in the anti-inflammatory and apoptotic effects of curcumin (diferuloylmethane). Free Radic. Biol. Med. 2007, 43(4), 568–580.

Saxena, G., Singh, M., Meena, P., Barber, S., Sharma, D., Shukla, S., Bhatnagar, M. Neuroprotective effects of Asparagus racemosus Linn Root Extract: An Experimental and Clinical Evidence. Annals of Neurosci. 2007 14, 3.

Scartezzini, P., Speroni, E. Review of some plants of Indian traditional medicine with antioxidant activity. J. Ethnopharmacol. 2000, 71, 23–43.

Schapira, A. H., Bezard, E., Brotchie, J. Novel pharmacological targets for the treatment of Parkinson's disease. Nat. Rev. Drug. Discov. 2006, 5, 845–854.

Sehgal, N., Gupta, A., Valli, R. K., Joshi, S. D., Mille, J. T., Hamel, E., Khanna, P., Jain, S. C., Thakur, S., Ravindranath, V. Withania somnifera reverses Alzheimer's disease pathology by enhancing low-density lipoprotein receptor-related protein in liver. PNAS. 2012, 109(9), 3510–3515.

Sharma, P. V. Ashwagandha, Dravyaguna Vijana, Chaukhambha Viashwabharti. Varanasi: 1999 pp. 763–765.

Sharma, H. M., Hanna, A., Kauffman, E. M., Newman, H. A. I. Inhibition of human low-density lipoprotein oxidation *in vitro* by Maharishi Ayurveda herbal mixtures. Pharmacol. Biochem. Behav. 1992, 43, 1175–1187.

Sharma, P. C., Yelne, N. B., Dennis, T. J. Data base on Medicinal Plants used in Ayuveda, CCRAS, New Delhi, 2001, Vol.1.

Sharma, R., Chaturvedi, C., Tewari, P. V. Efficacy of Bacopa monnieri in revitalizing intellectual functions in children. Indian J. Med. Res. 1987, 6, 1–10.

Sharma, S. K., Singh, A. P. *In vitro* antioxidant and free radical scavenging activity of Nardostachys jatamansi DC. J. Acupunct. Meridian Stud. 2012, 5(3), 112–118.

Shinomol, G. K., Muralidhara. Effect of Centella asiatica leaf powder on oxidative markers in brain regions of prepubertal mice in vivo and its *in vitro* efficacy to ameliorate 3-NPA-induced oxidative stress in mitochondria. Phytomedicine. 2008, 15, 971–984.

Shinomol, G. K., Mythri, R. B., Srinivas, B., Muralidhara, M. M. Bacopa monnieri extract offsets rotenone-induced cytotoxicity in dopaminergic cells and oxidative impairments in mice brain. Cell. Mol. Neurobiol. 2012, 32(3), 455–465.

Shinomol, G. K., Raghunath, N., Bharath, M., Muralidhara. Prophylaxis with Bacopa monnieri Attenuated Acrylamide Induced Neurotoxocity and Oxidative Damage via Elevated Antioxidant Function. Cent. Nerv. Syst. Agents Med. Chem. 2013 13(1), 3–12.

Shinomol, G. K., Ravikumar, H., Muralidhara. Prophylaxis with Centella asiatica confers protection to prepubertal mice against 3-nitropropionic-acid-induced oxidative stress in brain. Phytother. Res. 2010, 24, 885–892.

Shinomol, G. K. M. Prophylactic neuroprotective property of Centella asiatica against 3-nitropropionic acid induced oxidative stress and mitochondrial dysfunctions in brain regions of prepubertal mice. Neurotoxicology. 2008, 29, 948–957.

Siddique, Y. H., Ara, G., Beg, T., Faisal, M., Ahmad, M., Afzal, M. Antigenotoxic role of Centella asiatica L. extract against cyproterone acetate induced genotoxic damage in cultured human lymphocytes. Toxicol. *In Vitro*. 2008, 22, 10–17.

Simons, M., Keller, P., De Strooper, B., Beyreuther, K., Dotti, C. G., Simons, K. Cholesterol depletion inhibits the generation of beta-amyloid in hippocampal neurons. Proc. Natl. Acad. Sci. U. S. A. 1998, 95, 6460–6464.

Singh, G. K., Garabadu, D., Muruganandam, A. V., Joshi, V. K., Krishnamurthy, S. Antidepressant activity of Asparagus racemosus in rodent models, Pharmacol. Biochem. Behav. 2009, 91, 283–290.

Singh, N., Bhalla, M., Jager, P., Gilca, M. An Overview on Ashwagandha: A Rasayana (Rejuvenator of Ayurveda. Afr. J. Tradit. Complement. Altern. Med. 2011, 8(5 Suppl), 208–213.

Singh, R. H., Narsimhamurthy, K., Singh, G. Neuronutrient impact of Ayurvedic Rasayana therapy in brain aging. Biogerontology. 2008, 9, 369–374.

Smith, M. A., Sayre, L. M., Monnier, V. M., Perry, G. Radical Ageing in Alzheimer's disease. Trends Neurosci. 1995, 18(4), 172–176.

Smith, P. F., Maclennan, K., Darlington, C. L. The neuroprotective properties of the Ginkgo biloba leaf: a review of the possible relationship to platelet-activating factor (PAF). J. Ethnopharmacol. 1996, 50, 131–139.

Solomon, P. R., Adams, F., Silver, A., Zimmer, J., DeVeaux, R. Ginkgo for memory enhancement: a randomized controlled trial. JAMA. 2002, 288, 835–840.

Soumyanath, A., Zhong, Y., Henson, E., Wadsworth, T., Bishop, J., Gold, B., Quinn, J. Centella asiatica Extract Improves Behavioural Deficits in a Mouse Model of Alzheimer's Disease: Investigation of a Possible Mechanism of Action. Intl. J. Alz. Dis. 2012, 3, 819–874.

Stridh, M. H., Correa, F., Nodin, C. Enhanced glutathione efflux from astrocytes in culture by low extracellular Ca(2+) and curcumin. Neurochem. Res. 2010, 35, 1231–1238.

Subathra, M., Shila, S., Devi, S. M., Panneerselvam, C. Emerging role of Centella asiatica in improving age-related neurological antioxidant status. Exp. Gerontol. 2005, 40, 707–15.

Takeda, A., Smith, M. A., Avila, J. In Alzheimer's disease, heme oxygenase is coincident with Alz50, an epitope of τ induced by 4-hydroxy-2-nonenal modification. J. Neurochem. 2000, 75(3), 1234–1241.

Thompson, P. M., Vinters, H. V. Pathologic lesions in neurodegenerative diseases. Prog. Mol. Biol. Transl. Sci. 2012, 107, 1–40.

Tohda, C., Kuboyama, T., Komatsu K. Dendrite extension by methanol extract of Ashwagandha (roots of Withania somnifera) in SK-N-SH cells. Neuroreport. 2000, 11, 1981–85.

Tohda, C., Kuboyama, T., Komatsu, K. Search for natural products related to regeneration of the neuronal network. Neurosignals. 2005, 14, 34–45.

Toniolo, R., Narda, F., Susmel, S., Martelli, M., Martelli, L., Bontempelli, G. Quenching of superoxide ions by curcumin. A mechanistic study in acetonitrile. Ann Chim. 2002, 92, 281–288.

Tsang, A. H., Chung, K. K. Oxidative and nitrosative stress in Parkinson's disease. Biochim. Biophys. Acta. 2009, 1792, 643–60.

Tundo, G., Ciaccio, C., Sbardella, D., Boraso, M., Viviani, B., Coletta, M., Marini. Somatostatin modulates insulin-degrading-enzyme metabolism: implications for the regulation of microglia activity in AD. PLoS One. 2012, 7:e34376.

Uabundit, N., Wattanathorn, J., Mucimapura, S., Ingkaninan, K. Cognitive enhancement and neuroprotective effects of Bacopa monnieri in Alzheimer's disease model. J. Ethnopharmcol. 2010, 127, 26–31.

Uniyal, M. R., Issar, R. K. Commercially and traditionally important medicinal plants of Mandakini Valley of Uttarakhand, Himalayas. J. Res. Indian. Med. 1969, 4, 83.

Vayda, A. B., Rajgopalan, T. S., Mankodi, N. A. Treatment of Parkinson's disease with the cowhage plant - Mucuna pruriens (Bak). Neurol. India. 1978, 36, 171–6.

Veerendra, Kumar, M. H., Gupta, Y. K. Effect of different extracts of Centella asiatica on cognition and markers of oxidative stress in rats. J. Ethnopharmacol. 2002, 79, 253–260.

Veerendra, S. G., Kumar, M. H., Gupta, Y. K. Effect of Centella asiatica on cognition and oxidative stress in an intracerebroventricular streptozotocin model of Alzheimer's disease in rats. Clin. Exp. Pharmacol. Physiol. 2003, 30, 336–342.

Vellas, B., Coley, N., Ousset, P. Long-term use of standardized ginkgo biloba extract for the prevention of Alzheimer's disease (GuidAge): a randomized placebo-controlled trial. Lancet Neurol. 2012, 11: 851–859.

Ven Murthy, M. R., Ranjekar, P. K., Ramassamy, C., Deshpande, M. Scientific basis for the use of Indian Ayurvedic medicinal plants in treatment of neurodegenerative disorders: ashwagandha. Cent. Nerv. Syst. Agents Med. Chem. 2010, 10(3), 238–246.

Vimal, S., Sissodia, S. S., Meena, P., Barber, S., Shukla, S., Saxena, A., Patro, N., Patro, I., Bhatnagar, M. Antioxidant effects of asparagus racemosus wild and Withania somnifera dunal in rat brain. Ann. Neurosci. 2010, 12, 67–70.

Visavadiya, N. P., Narasimhacharya, R. L. Hypolipidemic and antioxidant activities in Asparagus racemosus in hypercholesteremic rats. Indian Journal of Pharmacology. 2005, 37, 376–380.

Visweswari, G., Prasad, K., Chetan, P., Lokanath, V., Rajendra, W. Evaluation of the anticonvulsant effect of Centella asiatica (gotu kola) in pentylenetetrazol induced seizures with respect to cholinergic neurotransmission. Epilepsy and Behav. 2010, 17, 332–335.

Walker, A., G. The prefrontal cortex system in the R6/2 mouse model of Huntington's disease, Department of Psychological and Brain Sciences Vol. PhD, Indiana University p. 12, 2010.

Wang, J., Du, X. X., Jiang, H., Xi, J. X. Curcumin attenuates 6-hydroxydopamine-induced cytotoxicity by anti-oxidation and nuclear factor-kappa B modulation in MES23.5 cells. Biochem. Pharmacol. 2009, 78, 178–183.

Wang, M. S., Boddapati, S., Emadi, S., Sierks, M. R. Curcumin reduces alpha-synuclein induced cytotoxicity in Parkinson's disease cell model. BMC Neurosci. 2010, 11, 57.

Warrier, P. K., Nambiar, V. P. K., Ramankutty, C. In: Indian Medicinal Plants a Compendium of 500 species, Hyderabad, India, Orient Longman, 2000, Vol 5.

Wattanathorn, J., Mator, L., Muchimapura, S., Tongun, T., Pasuriwong, O., Piyawatkul, N. Positive modulation of cognition and mood in the healthy elderly volunteer following the administration of Centella asiatica. J. Ethnopharmacol. 2008, 116, 325–332.

White, K. G., Ruske, A. C. Memory deficits in Alzheimer's disease: the encoding hypothesis and cholinergic function. Psychonomic Bulletin and Review. 2002, 9, 426–437.

Widodo, N., Priyandoko, D., Shah, N., Wadhwa, R., Kaul, S. C. Selective killing of cancer cells by Ashwagandha leaf extract and its component Withanone involved ROS signaling. PLoS One. 2010, 5(10), e13536.

Wu, W. R., Zhu, X. Z. Involvement of monoamine oxidase inhibition in neuroprotective and neurorestorative effects of Ginkgo biloba extract against MPTP-induced nigrostriatal dopaminergic toxicity in C57 mice. Life Sciences 1999, 65(2), 157–164.

Xu, C., Qu, R., Zhang, J., Li, L., Ma, S. Neuroprotective effects of madecassoside in early stage of Parkinson's disease induced by MPTP in rats. Fitoterapia. 2013, 90, 112–118.

Xu, C., Wang, Q., Sun, L., I, X., Deng, J., Li, L., Zhang, L, Xu, R., Ma, S. Asiaticoside: Attenuation of neurotoxcitiy induced by MPTP in a rat model of Parkinsonism via maintaining redox balance and up regulating the ratio of Bcl2/Bax. Pharmcol. Biochem. Behav. 2011, 100, 413–418.

Xu, Y., Cao, Z., Khan, I., Luo, Y. Gotu Kola (Centella asiatica) extract enhances phosphorylation of cyclic AMP response element binding protein in neuroblastoma cells expressing amyloid beta peptide. J. Alzheimer's Dis. 2008, 13, 341–349.

Yadav, S., Prakash, J., Chouhan, S., Singh, S. P. Mucuna pruriens seeds extract reduces oxidative stress in nigrostriatal tissue and improves neurobehavioural activity in paraquat-induced Parkinsonian mouse model. Neurochem. Int. 2013, 62(8), 1039–1047.

Yang, F., Lim, G. P., Begum, A. N. Curcumin inhibits formation of amyloid β oligomers and fibrils, binds plaques, and reduces amyloid in vivo. J. Biol. Chem. 2005, 280(7), 5892–5901.

Yang, S. F., Wu, Q., Sun, A. S., Huang, X. N., Shi, J. S. Protective effect and mechanism of Ginkgo biloba leaf extracts for Parkinson disease induced by 1-methyl-4- phenyl-1,2,3,6-tetrahydropyridine. Acta. Pharmacol. Sin. 2001, 22

Yeh, Y. C., Liu, T. J., Wang, L. C., Lee, H. W., Ting, C. T., Lee, W. L. A standardized extract of Ginkgo biloba suppresses doxorubicin-induced oxidative stress and p53-mediated mitochondrial apoptosis in rat testes. Br. J. Pharmacol. 2009, 156, 48–61.

Yoritaka, A., Hattori, N., Uchida, K., Tanaka, M., Stadtman, E. R., Mizuno, Y. Immunohistochemical detection of 4-hydroxynonenal protein adducts in Parkinson disease. Proc. Natl. Acad. Sci. USA. 1996, 93, 2696–2701.

Zainol, M. K., Abd-Hamid, A., Yusof, S., Muse, R. Antioxidant activity and total phenolic compounds of leaf, root and petiole of four accessions of Centella asiatica (L) Urban. Food Chem. 2003, 81, 575–581.

Zandi, P. P., Anthony, J. C., Khachaturian, A. S., Stone, S. V., Gustafson, D., Tschanz, J. T., Norton, M. C., Welsh-Bohmer, K. A., Breitner, J. C., Cache County Study Group. Reduced risk of Alzheimer disease in users of antioxidant vitamin supplements: the Cache County Study. Archives in Neurology. 2004, 61, 82–88.

Zhang, L. L., Wang, H. S., Yao, Q. Q., Luan, Y., Wang, X. L. Determination of asiaticoside and madecassoside in Centella asiatica (L.) Urb by RP-HPLC. Chinese Traditional and Herbal Drugs. 2007, 38, 455–456.

CHAPTER 2

ALTERNATIVE AND COMPLEMENTARY MEDICINE IN TREATING FUNGAL DERMATOPHYTIC INFECTIONS

M. SHARANYA and R. SATHISHKUMAR

CONTENTS

ABSTRACT

Fungal skin infections caused by dermatophytes are a great menace and the existing medications possess several drawbacks, which overlay the way for seeking alternative and complementary medicine. Western Ghats is one of the hotspot of India and it is a rich source of flora and fauna. In this chapter, the medicinal plants, those are scientifically proved are accounted, the details of experimental methodology used, type of extraction, etc. is included to give the readers a complete overview of the present research towards the treatment of dermatophytosis. Focusing on the treatment with medicinal plant preparation will provide an alternative to overcome the prevailing difficulties in medical practice.

2.1 INTRODUCTION

Fungi are a group of eukaryotes, which are unicellular or multicellular, or syncytial spore-producing organisms with approximate size ranging from 2–10 μm, and may be either beneficial or harmful. The fungal cell contains membrane bound organelles like nuclei, mitochondria, golgi apparatus, endoplasmic reticulum, lysosomes, etc. and remarkably ergosterols in the external membrane forming a rigid structure with the help of chitin molecules. Fungi possess 80S ribosomes and the cell division exhibit mitosis. Basic requirement of fungi includes water, and oxygen, and perhaps fungi are said as chemoheterotrophs (require organic compounds for both carbon and energy sources), osmotrophic (obtain nutrients by absorption), saprophytes (lives on decaying matter) or parasites (lives on living matter) and there are no obligate anaerobes. In general, lipids and glycogen are the storage form and the reproduction occurs either asexually and/or sexually.

Living organisms are classified into five kingdoms namely Monera, Protista, Fungi, Plantae and Animalia where the fungi are placed in a separate kingdom by R.H. Whittaker in 1969. Fungi are classified based on sexual/asexual reproduction and morphology. The fungi are grouped as zygomycetes (produce zygospore), ascomycetes (produce endogenous spores called ascospores in cells called asci), basidiomycetes (produce exogenous spores called basidiospores in cells called basidia) and deuteromycetes (a heterogeneous group of fungi where no sexual reproduction has yet been demonstrated and also called fungi imperfecti). Similarly on morphological basis, they are moulds (e.g., *Aspergillus* spp., *Microsporum gypseum*, etc.), yeasts (e.g., *Cryptococcus neoformans*, *Saccharomyces cerviceae*), yeast like (e.g., *Candida albicans*) and dimorphic (e.g., *Histoplasma capsulatum*, *Blastomyces dermatidis*, *Paracoccidiodes brasiliensis*, *Coccidioides immitis*).

2.2 FUNGAL INFECTIONS

Among an estimated 1.5 million species of fungi, some 200 have been recognized as "human pathogens" and the infections are normally categorized as follows:

- Superficial (superficial phaeohyphomycosis, tinea versicolor, black piedra and white piedra)
- Cutaneous (dermatophytosis and dermatomycosis)
- Sub-cutaneous (chromoblastomycosis, rhinosporidiasis, mycetoma, sporotrichosis, subcutaneous phaeohyphomycosis, lobmycosis)
- Systemic (blastomycosis, histoplasmosis, coccidioidomycosis, paracoccidioidomycosis)
- Opportunistic (candidiasis, cryptococcosis, aspergillosis)
- Other mycoses (otomycosis and occulomycosis)
- Fungal allergies (asthma and sinusitis)
- Mycetism and Mycotoxicosis (anorexia, oedema of legs, massive gastrointestinal bleeding)

The infections are mostly diagnosed by specimen collection and further followed by microscopy or using other techniques such as culturing, serology, antigen detection, skin tests and molecular techniques accordingly to the type and site of infection.

2.2.1 FUNGAL SKIN INFECTION

Various microscopic organisms live harmlessly inside the body and on the surface of the skin. However, certain types of fungus that are normally harmless, on overgrowth can cause superficial and systemic infections which are more commonly seen in those people undertaking antibiotics, corticosteroids, immunosuppressant drugs and contraceptives. Even also prevail in people with endocrine disorders, immune diseases, diabetes and others such as AIDS, tuberculosis, major burns and leukemia. Moreover, found in obese people with excessive skin folds (Common Fungal Infections of the Skin, Spring 2006). In skin infections the topmost layer called stratum corneum is highly affected especially when exposed to a barrage of insults from the environment and thrive mostly on moist areas of the body, such as under the breasts, in the groin, and between fingers and toes. Some may cause no discomfort but others involve itching, swelling, and pain. Cutaneous fungal infections are often divided into 'superficial' and 'deep' forms (Kovacs and Hruze, 1995). The most common types of superficial infections are ringworm, athlete's foot and jock itch, and are approximately 25% of the populations afflicted with them (Common Fungal Infections of the Skin, Spring 2006). In the case of inflammatory conditions the fungi multiply and invade the skin, digestive tract, the genitals, or other body tissues. The most common fungal infections that infect the skin belong to a class of fungus called "tinea". Tinea refers exclusively to dermatophyte infections occur on skin, hair and nails (Continuum of Care, Judith Stevens). Noble and Forbes (1998) quoted "among the most common skin diseases, specifically superficial fungal infections affect millions of people around the world".

2.2.1.1 DERMATOPHYTES

The dermatophytes are capable of invading keratinized tissues of humans and other animals to cause an infection, dermatophytosis, commonly referred to as ringworm but are not able to penetrate deeper tissues or organs of immunocompetent hosts (Weitzman and Summerbell, 1995). Dermatophytes exist as both keratinophilic and keratinolytic and are often live on dead tissues. The infection may be mild to severe according to the host immune condition and reaction to the metabolic products of the fungus, the virulence of the infecting strain or species, the anatomic location of the infection and the local environmental factors. Dermatophytes as saprophytes reproduce asexually *via* sporulation of anthro-, micro- and macroconidia produced from the specialized conidigenous cells. Vegetative structures are observed with typical arrangement on hyphae, chlamydospores, spirals, antler-shaped hyphae (chandeliers), nodular organs, pectinate organs and racquet hyphae. Growth forms and pigmentation produced by the dermatophyte colonies would be the presumptive identification of the species where five most important colony characteristics had been listed by Ajello (1966), they are: (1) rate of growth (2) general topography (flat, heaped, regularly or irregularly folded) (3) texture (yeast-like, glabrous, powdery, granular, velvety or cottony) (4) surface pigmentation, and (5) reverse pigmentation.

According to WHO (World Health Organization), the dermatophytes are defined in three genera: *Epidermophyton*, *Trichophyton* and *Microsporum* (Palacio et al., 2000). They comprise about 40 different species, and have common characteristics:

1. Close taxonomic relationships;
2. Keratinolytic properties (they all have the ability to invade and digest the keratin as saprophytes "*in vitro*" and as parasites "*in vivo*", producing lesions in the living host);
3. Occurrence as etiologic agents of infectious diseases of man and/or animals.

Ecologically dermatophytes are grouped as:

i) zoophilic Keratin utilizing on hosts–found in living animals (e.g., *M.canis*, *T.verrucosum*);
ii) anthropophilic Keratin utilizing on hosts–found in humans (e.g., *M.audounii*, *T.tonsurans*);
iii) geophilic Keratin utilizing soil saprophytes (e.g., *M.gypseum*, *T.ajelloi*).

The dermatophytic infections are named according to the site of infection (Table 2.1).

TABLE 2.1 Nosology of Dermatophyte Infections

S. No.	Infection Name	Area of infection	Organism	Description
1.	Tinea capitis	Scalp infection	*M.canis, M.audounii, T.tonsurans, T.verrucosum*	Small papule spreads to form scaly, irregular or well-demarcated areas of alopecia. The cervical and occipital lymph nodes may be enlarged; a kerion, a boggy, inflammatory mass, followed by healing.
2.	Tinea corporis	Trunk and limb infection	*T.rubrum, T.verrucosum, M.canis*	Single or multiple scaly annular lesions with a slightly elevated, scaly and or erythematous edge, sharp margin and central clearing.
3.	Tinea barbae	Hair infection	Zoophilic, *Trichophyton* sp.	Patches of inflammation, sometimes with follicular pustules in beard area
4.	Tinea faciei	Facial infection	Zoophilic, *Trichophyton* sp.	Patches of inflammation, sometimes with follicular pustules in surface of the face
5.	Tinea cruris	Groin infection	*T.rubrum, T.interdigitale, E.floccosum*	Most common in men, typical lesion is red, marginated eruption which spreads outwards from groin crease with edges may scaly, pustular or vesicular
6.	Tinea pedis	Foot infection	*T.rubrum, T.interdigitale, E.floccosum*	Occur in three distinct forms; interdigital toe webs fissured, macerated and itchy, vesicular patches affect soles and sides of feet lead to blisters and itchy, dry scaly changes over whole plantar surface and extending up the sides of the feet, producing a demarcated line (moccasin pattern)
7.	Tinea manuum	Hand infection	*T.rubrum* (rarely geophilic may also occur)	Unilateral, diffuse scaling of the palm.
8.	Tinea unguium	Nail infection	*T.rubrum, T.interdigitale*	Mostly toe-nail infected, nails separate from the nail-bed, the nail plate thickens and crumbly and yellow-brown.

2.2.1.2 CURRENT TREATMENTS

The infection by dermatophytes shows acute and chronic inflammatory changes in the dermis and can be treated with the antifungal agents either by topical application or by oral intake. Topical agents should posses the ability to penetrate the stratum corneum cells whereas the oral treatment would be suitable in the case of inflammatory infections and hyperkeratotic lesions (Palacio et al., 2000).

The existing classes of antifungal agents are polyenes, azoles and pyrimidines, others include allylamines, candins and the drug griseofulvin each targets the fungal cell in their unique mode of action (see Figure 2.1). Flucytosine, belonging to the pyrimidine class of antifungal agent, is the only drug targeting the thymidylate synthetase and thereby interrupt in the DNA synthesis (Vermes et al., 2000). On long usage, flucytosine exhibit toxicity to bone marrow which further leads to anemia, leucopoenia, thrombocytopenia and also causes nausea, vomiting, diarrhea, liver damage, nephrotoxicity, abdominal cramps and pain. Allylamines and azoles inhibit the ergosterol biosynthesis pathway where allylamines specifically reduces the squalene oxide formation and the azoles on 14-α-demethylase thereby facilitating the accumulation of lanosterol (Ghannoum et al., 1999). Use of azoles and allylamines for long time may cause hepatitis, teratogenic, allergic rash, hormone imbalance, nausea, vomiting and fluid retention. Polyenes disrupt the fungal cell membrane and pave the way for cell content leakage. However, prolong usage would result in infusion related events such as fever, chills, headache, nausea, vomiting and dose-limiting nephrotoxicity (Gallis et al., 1990). Echinocandin, is the only available semi synthetic class of drug in the market is also known as fungal (1,3)-β-D-glucan synthase inhibitors which associate in disruption of cell wall biosynthesis (Current et al., 1995).

2.3 MEDICINAL PLANTS AS ALTERNATIVE AND COMPLEMENTARY MEDICINE

Medicinal plants play a prominent role in treating most number of diseases. The World Health Organization (WHO) estimates that over 80 percent of people in developing countries depend upon traditional medicine for treatment of diseases and woe in their primary health care (Srivastav

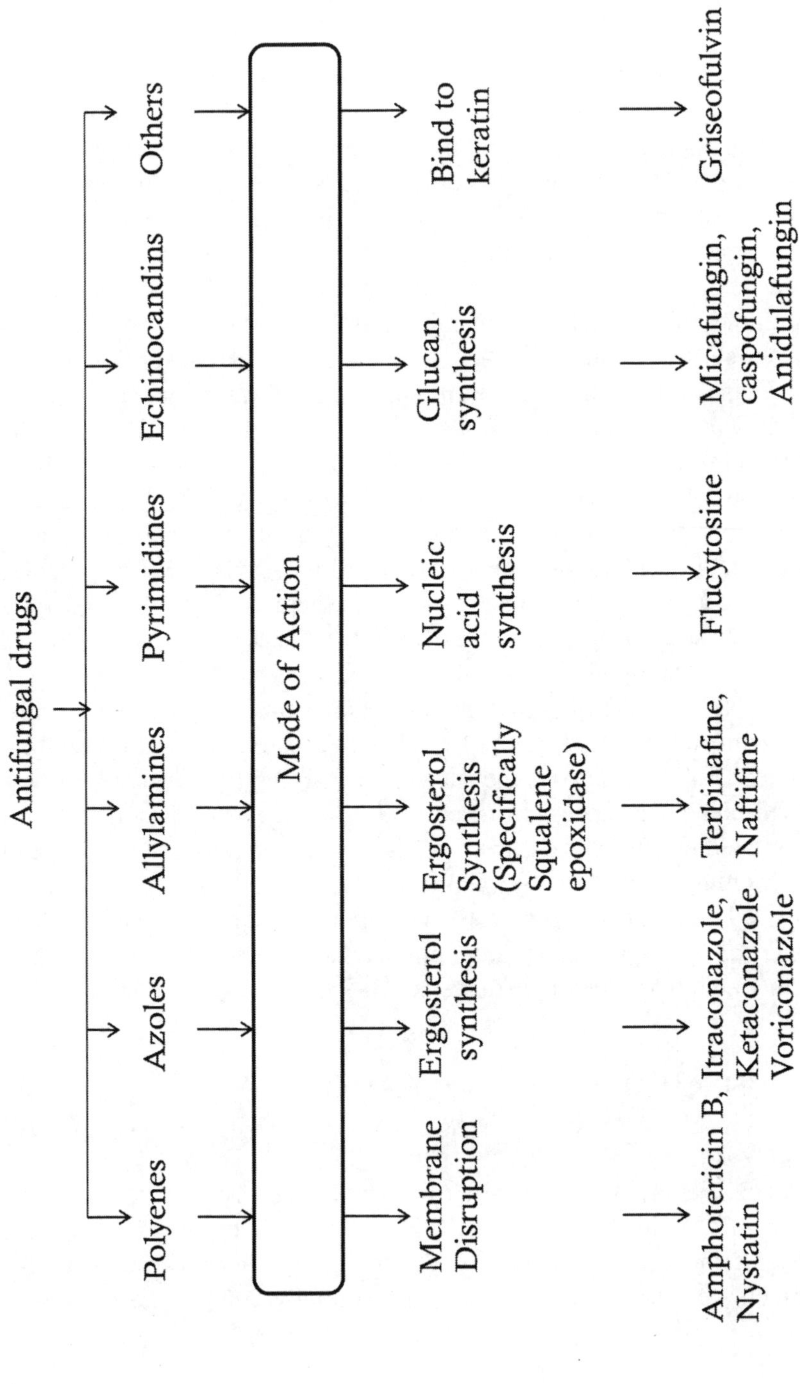

FIGURE 1 Classes of antifungal agents and its mode of action

et.al., 2011). Since the existing drugs are determined with side effects and are also be reasoned for reoccurrence of the infection either at same or at different site of the body and in the development of fungal resistance. Therefore, alternative and complementary medication to treat dermatophytosis is mandatory. India, is a rich source of medicinal plants in which Western Ghats is considered as one of the hotspot of the country. At present, about 40% or 60,000 sq. km of the Western Ghats is declared as an Ecologically Sensitive Area (ESA). According to the survey conducted by Krishnan et al. (2011) one third of the plant species are being endemic and about 500 species are categorized with medicinal importance and has put forth several measures in the flora and fauna conservation. A project conducted in the period of 2005–2008 had surveyed and recorded the total number of medicinal plants along with their botanical details from the area of Western Ghat, has been considered and the number of plants reported scientifically for its anti-dermatophytic activity are accounted for its possible usage as an alternate and complementary medicine (Project by Kholkute, 2005–2008, submitted to ICMR).

2.3.1 WESTERN GHATS

The biogeographic zone of Western Ghats includes a narrow stretch along India's west coast approximately 30–50 km inland, starting from the hills south of the Tapi river in the north to Kanyakumari in the south. Western Ghats is otherwise called as "Sahyadri" traverse the States of Kerala, Tamil Nadu, Karnataka, Goa, Maharashtra and Gujarat (see Figure 2.2). These mountains cover an area of around 140,000 km^2 in a 1,600 km long stretch that is interrupted only by the 30 km Palghat Gap at around 11°N. The Ghats rise up abruptly in the west to a highly dissected plateau up to 2900 m in height, and descend to dry Deccan plains below 500 m in the east. The extreme climatic and altitudinal gradient has resulted in a variety of forest types, from evergreen to semi-evergreen, from moist deciduous to dry deciduous formations. According to some biogeographers (Rao and Sagar, 2012) the Western Ghats forms the "Malabar" province and are internationally recognized as a region of immense global importance for its outstanding features and enormous biodiversity of ancient lineage. Although the Western Ghats cover only 5% of India's total geographical area, the region contains over 30% of the country's plant species of which around 12,000 species, from lower groups to flowering plants, are estimated to occur here.

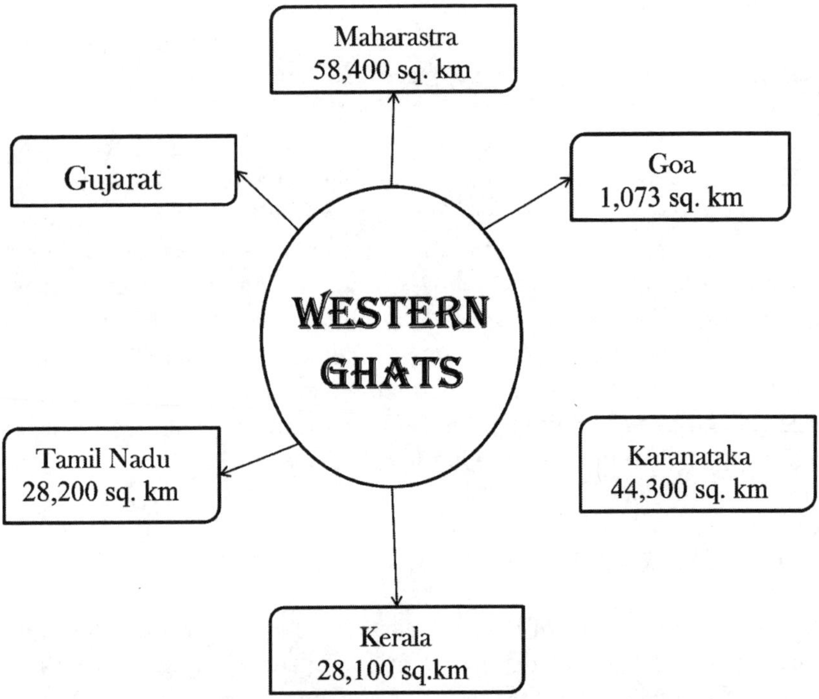

FIGURE 2.2 Area under Western Ghats.

2.3.1.1 MEDICINAL PLANTS REPORTED WITH ANTIDERMATOPHYTIC ACTIVITY

2.3.1.1.1 Acacia mearnsii

Acacia mearnsii De Wild. (Fabaceae) is found in abundance throughout Australia, Asia, Africa and America. The plant was known previously as *Racosperma mearnsii* and commonly known as Black Wattle tree, it is a short medium lived woody perennial and spreading tree, about 15 m high with smooth and greenish-brown bark on young branches which gradually turn blackish and rough on trunk. The young branchlets are downy. While it is widespread and common in lowlands, open forest, healthy woodland and on cleared land, particularly on dry, shallow soils (Walsh and

Entwisle, 1996), it grows in open forest, woodland or tussock grassland, in gullies or on hillsides, and in sandy or gravelly clay soils. It has a globular inflorescence with 20–30 tiny pale yellow flowers. Pods, dark brown to black in color, are more or less straight, 5–10 cm long, 5–8 mm wide and strongly constricted between seeds (Maslin, 2001). The seeds are reported for its antimicrobial activity where the minimal inhibitory concentration (MIC) and minimal fungicidal concentration (MFC) lies at 2500 μg/mL and 5000 μg/mL against both *T. mucoides* and *T. tonsurans* (Olufunmiso et al., 2012).

2.3.1.1.2 Acalypha indica

Acalypha indica L. belongs to Euphorbiaceae family, is an annual, erect herb, up to 1 m high. Leaves are 2.5–7.5 cm long, ovate or rhomboid-ovate, crenate-serrate. Flowers in numerous lax, erect, elongated axillary spikes, the male minute, clustered near the summit of the spike, the females scattered, surrounded by a large. Traditionally the leaves mixed with common salt is applied to scabies and other skin diseases (Indian Herbal Remedies, Khare CP). The leaves extracted with water and ethanol showed activity against *T.rubrum* (water: MIC-9.3 μg/mL, MFC-9.3 μg/mL; ethanol: MIC-9.3 μg/mL, MFC-9.3 μg/mL), *T.mentagrophytes* (water:MIC-9.3 μg/mL, MFC-9.3 μg/mL; ethanol: MIC-9.3 μg/mL, MFC-9.3 μg/mL) with respective MIC and MFC values (Vaijayanthimala et al., 2004). The study by Radhika et al. (2013) also revealed the significant activity exhibited by the ethanol and ethyl acetate extract against *T.rubrum*, *T.mentagrophytes*, *M.gypseum* and *T.tonsurans* showing MIC and MFC at 250 μg/mL for all the organisms rather than the hexanic extract which exhibited at higher 1000 μg/mL concentration.

2.3.1.1.3 Achyranthes aspera

Achyranthes aspera L. (Amaranthaceae) is commonly known as Rough Chaff tree is an erect or procumbent, annual or perennial herb, 1–2 m in height, often with a woody base, commonly found as a weed of waysides, on roadsides. Stems angular, ribbed, simple or branched from the base, often with tinged purple color, branches terete or absolutely quadrangular, striate, pubescent, leaves thick, 3.8–6.3 × 22.5–4.5 cm, ovate- elliptic or obovate-rounded, finely and softly pubescent on both sides, entire, petiolate, petiole 6–20 mm long, flowers greenish white, numerous in axillary or terminal spikes up to 7 5 cm long, seeds sub-cylindric, truncate at the apex, rounded at the base, reddish brown (Srivastav et al., 2011). The plant is used in treating skin diseases especially its oil relieves all kind of skin problems and the plants is also called prickly flower (http://www.astrogle.com/). The method of extraction also had impact on the plants biological activity. Study by Londonkar et al. (2011) depicts the scenario where the leaves extracted with chloroform, petroleum ether and methanol using infusion and maceration methods were evaluated for its anti-dermatophytic activity. Where the methanol extract obtained by infusion method has shown more significant inhibitory activity against the dermatophytes *T.rubrum* and *M.canis* with 14 mm and 12 mm dia of zone of inhibition for *T.rubrum* and 12 mm dia for *M.canis*.

Acorus calamus

Acorus calamus L. (Araceae), which is commonly known as sweet flag, is a herbaceous perennial with a rhizome that is long indefinite branched, smooth, pinkish or pale green. Its leaf scars are brown white and spongy and it possess slight slender roots. The leaves are few and distichously alternate whose size was found to be between 0.7–1.7 cm wide with average of 1 cm. The flowers are 3 to 8 cm long, cylindrical, greenish brown and contains multitude of rounded spikes covering it. The fruits are found to be small and berry like with few seeds (Balakumbahan et al., 2010). Traditionally the plant is used as a promising immunomodulatory agent in the inflammatory skin diseases (Divya et al., 2011) besides novel terpenoid, 1–2, 4, 5 trimethoxy phenyl 1' methoxy propionaldehyde (TMPMP) isolated was scientifically tested and reported to cure tinea pedis in wistar rat model infected with *T.rubrum* (Subha and Gnanamani, 2011).

2.3.1.1.4 Aegle marmelos

Aegle marmelos L. Correa (Rutaceae) is commonly called Vilvam in Tamil, Bael, Bengal quince or stone apple and are often cultivated in temples for its leaves which are used in poojas. The leaves, stem, bark and fruits of this plant have long been used in traditional medicine. Bael is a slow growing, tough subtropical tree and is the only plant belonging to the genus Aegle (Sharma et al., 2007). The tree grows wild in well-drained soil and attains a size of about 12 to 15 m height even in the harsh and dry climates. The branches contain spines that are arrow and are upto an inch in size. The leaves are alternate borne singly or in twos or threes and are made up of three to five oval, pointed, shallowly toothed leaflets. The flowers are fragrant and are found in clusters along the young branches (Maity et al., 2009). The plants are rich source of bioactive compounds, each exist with its unique activity. Besides, quantity of the compounds varies conceivably from plant to plant and even within the parts of the same plant. In such a way, leaves were collected, dried and subjected to cold extraction with water and 100% ethyl alcohol and the coarse powder was also successively extracted with various organic solvents like hexane, benzene, chloroform, ethylacetate, methanol and water. Different fractions were collected, filtered and evaporated to dryness and are further evaluated for its antifungal activity on dermatophytes such as *T.mentagrophytes*, *T.rubrum*, *M.canis*, *M.gypseum* and *E.floccosum*. MIC and MFC values observed for all the extractions were at 400 µg/mL concentration beneath the methanol fraction, ethanol and water extract exhibited at 200 µg/mL against *T.mentagrophytes*, *M.canis* and *E.floccosum* (Balakumar et al., 2011). The susceptibility of every organism varies according to the type of organic solvents used for plant extraction. Therefore, attention has to be paid in remitting the crude plant extracts or plant compounds into the suitable medication.

2.3.1.1.5 *Aloe vera*

Aloe vera L. Burm. f. belongs to the family Liliaceae, is a hardy perennial plant with turgid green leaves joined at the stem in a rosette pattern. Leaves are formed by a thick epidermis (skin) covered with cuticle surrounding the mesophyll. It has been used externally to treat various skin conditions such as cuts, burns and eczema. It is said to be a miracle plant because of its medicinal and traditional uses (Rajeswari et al., 2012). Mosunmola et al. (2013) carried out antidermatophytic activity of *Aloe vera* juice using two methods such as agar disc and agar well diffusion and concluded that both the testing methods were reliable and exert reproducibility. The determined Minimum Inhibitory Dilution (MID) and Minimum Fungicidal Dilution (MFD) revealed the toxicity possessed by the juice obtained from different areas (Ilorin and Shagamu, Nigeria) against *E.floccosum*, *M.audouinii*, *T.mentagrophytes*, *T.rubrum*, *T.schoenleinii* and *M.ferrugineum*. The efficiency in curing dermatophytosis improved when associated with garlic pills in treating equines infected with *T.equinum* (Ferdowsi et al., 2012). The study revealed the progressive improvement on every five-day observation and a desirable response was observed at the end of 25th day.

2.3.1.1.6 *Alpinia galanga*

Alpinia galanga Willd. (Family-Zingiberaceae) is used in culinary, medication and cosmetics for centuries. It is commonly known as Rasna and Sugandhmula in Sanskrit and Arattai in Tamil. It is a perennial herb found commonly throughout the Western Ghats, Mysore, Goa, Malabar and Gujarat. Roots are adventitious, in groups, fibrous, persistent in dried rhizomes, about 0.5 to 2 cm long and 0.1 to 0.2 cm in diameter and yellowish brown in color. Rhizomes are cylindrical, branched, 2 to 8 cm in diameter, longitudinally ridged with prominent rounded warts (remnants of roots) marked with fine annulations; scaly leaves arranged circularly, externally reddish brown, internally orange yellow; odor pleasant and aromatic; spicy and sweet in taste (Chudiwal et al., 2010). In 1985, Janssen and Scheffer found terpinen-4-ol is the most active compound from the essential oils of fresh and dried rhizomes of *A.galanga* and also identified the presence of acetoxychavicol acetate through mass spectroscopy (MS) and nuclear magnetic resonance (NMR). In addition, the compound showed significant inhibitory activity against tested dermatophytes with MIC value ranging from 50 to 250 μg/mL. The 95% ethanolic extract of *A.galanga* rhizome exhibited inhibitory action on the growth of *M.canis*, *M.gypseum*, *T.mentagrophytes* (Trakranrungsie et al., 2008).

2.3.1.1.7 *Alangium salvifolium*

Alangium salvifolium (L.f) Wang. belongs to the family Alangiaceae and locally called as Ankolam (Tariqo and Javed, 1985). The plant is distributed in dry regions, plains and lower hills in India, their roots are useful for external application in acute case of rheumatism, leprosy and inflammation and internal application in cases of bites of rabbit and dogs (Bakhru, 1997). Water extract of the ground wood was evaluated for its antidermatophytic activity and it has been concluded that the plant can be included in the herbal preparation for the treatment of some dermatomycotic infections (Mansuang et al., 2002).

2.3.1.1.8 *Andrographis paniculata*

Andrographis paniculata Nees (Family-Acanthaceae) is commonly known as 'King of Bitters'. It is the most popular and extensively used in Ayurveda, Unani and Siddha medicines and also as home remedy for various diseases in Indian traditional as well as tribal medicine. It is an annual, profusely branched, erect herb, and 0.5–1.0 m in height with a tap root. Leaves are green, lanceolate, 3–7 cm × 1–2.3 cm in size, glabrous with slightly undulate margin, acuminate apex with a tapering base. Flowers are small and solitary; corolla whitish or light pink in color with hairs. Fruit, a capsule, linear, oblong and acute at both ends; seeds numerous (Niranjan et al., 2010). The dermatophyte *E. floccosum* is more susceptible to ethanolic extract of *A. paniculata* leaves showing 74.6% of inhibition where as *T. rubrum* exhibited about 70.9% inhibition (Ramya and Lakshmi Devi, 2011).

2.3.1.1.9 *Anthocephalus indicus*

Anthocephalus indicus Rich. (Rubiaceae) is a large tree with a broad umbrella-shaped crown and straight cylindrical bole. The branches are characteristically arranged in tiers. The tree may reach a height of 45 m with a stem diameter of 100–160 cm and sometimes it has a small buttress up to 2 m high. The bark is grey, smooth and very light in young trees, but rough and longitudinally fissured in old trees. The branches spread horizontally and drop at the tip. The leaves are glossy green, opposite, simple sessile to petiolate, ovate to elliptical (15–50 cm long by 8–25 cm wide). The ethanolic and hot water extracts of the ripened fruit was found inhibiting the growth of *T.rubrum* at MIC value 2 mg/mL (Mishra and Siddique, 2011).

2.3.1.1.10 *Artemisia nilagarica*

Artemisia nilagarica (Clarke) Pampan. is commonly called as Indian worm wood, belongs to the family Asteraceae. It is an aromatic shrub, 1–2 m high, yellow or dark red small flowers, grows throughout India in hills up to 2400 m elevation. It is erect, hairy, often half-woody and the stems are leafy and branched. The leaves are pinnately lobed, 5–14 cm long, gray beneath. Mugwort blossoms with reddish brown or yellow flowers. The flowers are freely small and stand in long narrow clusters at the top of the stem. The fruit (achene) is minute. The essential oils from the leaves was determined with significant antidermatophytic activity against *T.rubrum*, *M.canis* and *M.gypseum* at 500 µg/mL, 125 µg/mL and 62.50 µg/mL MIC values and the measured zone of inhibition were 26 mm, 28 mm and 29 mm respectively (Vijayalakshmi et al., 2010).

2.3.1.1.11 *Asclepias curassavica*

Asclepias curassavica L. (Asclepiadaceae) is an erect perennial shrub, growing over a meter high; stems are green at the base and reddish at the top. Stems and leaves exude a milky sap, when damaged. Leaves are opposite, dark green to reddish green, long and narrow (6 –15 cm), tapering to a point at both ends, and located towards the end of the stems. Small bunches of flowers grow at the ends of branches. They are red, with an orange center, the petals are curved backwards. The fruit is a long narrow pod, which splits open to set loose flat seeds with silky hairs at one end. Hexanic and methanolic extract of leaves and stems exhibited activity against *T.mentagrophytes* and *T.rubrum* at MIC value >8.0 and 2.0 mg/mL concentration (Garcia et al., 2003) respectively. Among the few medicinal plants evaluated, hexane, ethylacetate and methanol revealed the significant activity against all the organisms, *T.mentagrophytes*, *T.simii* and *E.floccosum* except *T.rubrum* (Duraipandian and Ignacimuthu (2011).

2.3.1.1.12 *Azadirachta indica*

Azadirachta indica commonly known as neem (Meliaceae) is native of India and naturalized in most of tropical and subtropical countries are of great medicinal value and distributed widespread in the world. It is a tree 40–50 feet or higher, with a straight trunk and long spreading branches forming a broad round crown; it has rough dark brown bark with wide longitudinal fissures separated by flat ridges. The leaves are compound, imparipinnate, each comprising 5–15 leaflets. The compound leaves are themselves alternating with one another. It bears many flowered panicles, mostly in the leaf axils. The selel are ovate and about one cm long with sweet scented white oblanciolate petals. It produces yellow drupes that are ellipsoid and glabrous, 12–20 mm long. Fruits are green, turning yellow on ripening, aromatic with garlic like odour. Fresh leaves and flowers come in March-April. Fruits mature between April and August depending upon locality (Hashmat et al., 2012).

Each and every parts of the neem triumph as a cure in almost all kind of diseases. Henceforth, the extraction of leaves and stem using water, ethanol and all kind of organic solvents exhibit significant toxicity against the pathogens (Radhika and Michael, 2013) relative to the previous report showing least MIC value of about 0.57 μg/mL by the ethanolic extract of stem (Vaijayanthimala et al., 2004). The maximum effect was observed in the neem seed and leaf against the pathogens *T.rubrum* and *Candida albicans* which are isolated from the HIV +ve infected immune deficiency patients (Santosh kumar et al., 2012).

2.3.1.1.13 Bauuhinia variegate

Bauhinia variegate is commonly called orchid tree, is a semideciduous tree to 15 m (50 ft.) tall, with a spreading crown. Leaves alternate, long petioled (to 3 cm [1.25 in] long), thin-leathery, simple but deeply cleft at apex, forming 2 large rounded lobes; lower surfaces downy, especially at top of petiole; blades with 11–13 veins extending from heart-shaped or rounded base. Flowers showy, fragrant, in few-flowered clusters near stem tips, appearing during leaf fall (early spring); 5 petals, clawed, overlapping, pale magenta to indigo (occasionally white), with dark red and yellow also on upper petal; 5 stamens (rarely 6). Fruit a flat, oblong pod, to 30 cm (1 ft.) long, 10–15-seeded. *T.mentagrophytes* and *T.rubrum* were found susceptible to the extracts obtained from the leaves showing activity between 40–55% (Gunalan et al., 2011).

2.3.1.1.14 Boerhaavia diffusa

Boerhaavia diffusa (Nictaginaceae) is a perennial creeping weed, prostrate or ascending herb, up to 1 m long or more, having spreading branches. The roots are stout and fusiform with a woody root stock. The stem is prostrate, woody or succulent, cylindrical often purplish, hairy, and thickened at the nodes. Leaves are simple, thick, fleshy and hairy, arranged in unequal pairs, green and glabrous above and usually white underneath. Flowers are minute, subcapitate, and present in a group of 4–10 together in small bracteolate umbels, forming axillary and terminal panicles. These are hermaphrodite, pedicellate and white, pink or pinkish-red in color.

Two or three stamens are present and are slightly exerted. The stigma is peltate. The achene fruit is detachable, ovate, oblong, pubescent, five-ribbed and glandular, anthocarpous, and viscid on the ribs (Sahu et al., 2008). *B.diffusa* root extracts extracted with different solvents exerts varying degree of inhibitory action on the *Microsporum* species (Agrawal et al., 2004a; Agrawal et al., 2004b).

2.3.1.1.15 Cassia tora

Cassia tora Linn. (Family-Caesalpiniaceae) is distributed throughout India, Sri Lanka, West China and tropics. It is an annual herbaceous foetid herb, almost an under-shrub, up to 30–90 cm high, with pinnate leaves. Leaflets are in three pairs, opposite, obovate, oblong with oblique base and up to 10 cm long. Flowers are in pair in axils of leaves with five petals and pale yellow in color. Pods are somewhat flattened or four angled, 10 to 15 cm long and sickle shaped, hence commonly named as sickle-pod. The seeds are 30–50 in a pod, rhombohedral and gathered in autumn (Jain and Patil, 2010). The leaves were extracted with cold methanol and observed for the antifungal activity where it showed strong inhibition for the growth of *M.canis* with 10 mm zone of inhibition and also inhibited the *T.rubrum* and *T.mentagraphytes* (Adamu et al., 2006). The organic extract of leaves specifically extracted with petroleum ether showed activity against *T.mentagrophytes* and *E.floccosum* showing about 10 and 20 mm of zone of inhibition (Rath and Mohanty, 2013). Therefore the extract has to be analyzed for its bioactive compound responsible for the activity.

2.3.1.1.16 Costus speciosus

Costus speciosus (Koen.) Smith belongs to the family zingiberaceae, is a perennial rhizomatous herb with erect or spreading stems. Leaves are simple, smooth, persistent, spirally arranged around the trunk. The leaves are sub sessile and appear dark green in color, elliptic or obovate in shape. The inflorescence is a spike around 10 cms long with large bracts in sub terminal position. Flowers are white in color, 5–6 cm long with a cup-shaped labellum and crest yellow stamens. Fruit is capsule and red in color. Seeds are black, five in number with a white fleshy aril (Rani et al., 2012). Hexane extract of the plant and the isolated two compounds sesquiterpenoids showed good activity showing MIC values of 62.5 μg/mL, 31.25

μg/mL, and 125 μg/mL against *T.simii*, *T.rubrum* and *E.floccosum* respectively (Duraipandiyan et al., 2012).

2.3.1.1.17 *Cryptolepis buchanani*

Cryptolepis buchanani Roem & Schult. commonly known as jambupatra sariva in Sanskrit, it is a large evergreen lactiferous, woody climbing, perennial shrub common especially in deciduous forest of sub-Himalayan tracts, Bihar, Orissa, East Uttar Pradesh in Varanasi region (Sharma et al., 2012). The methanol and aqueous extracts were evaluated against the human dermatophytic fungi and demonstrated marked inhibitory activity against *T.rubrum* (Vinayaka et al., 2010).

2.3.1.1.18 *Jatropha cucas*

Jatropha cucas L. or physic nut, is a bush or small tree (up to 5 m height) and belongs to the Euphorbia family. It has thick glorious branchlets, a straight trunk and grey or reddish bark, masked by large white patches. Leaves are of length and width of 6 to 15 cm with shallow lobes and are arranged alternately. Leaves were used traditionally as a medicine in treating dermatophytic infections. It was found with potential activity against the clinically collected dermatophytes- *Trichophyton*, *Microsporum* and *Epidermophyton* species at five different concentrations of 250, 200, 150, 100 and 50 mg/mL. The effective minimum inhibitory concentration was observed between 19.95 to 79.43 mg/mL by the ethanol extract (Aniebo et al., 2012).

2.3.1.1.19 *Lawsonia inermis*

Lawsonia inermis L. is a much branched glabrous shrub or small tree (2 to 6 m in height). Leaves are small, opposite in arrangement along the branches, sub-sessile, about 1.5 to 5 cm long, 0.5 to 2 cm wide, greenish brown to dull green, elliptic to broadly lanceolate with entire margin, petiole short and glabrous and acute or obtuse apex with tapering base. Young branches are green in color and quadrangular which turn red with age. Bark is greyish brown, unarmed when young but branches of older trees are spine tipped. Inflorescence is a large pyramid shaped cyme. Flowers are small, about 1 cm across, numerous, fragrant, white or rose colored with four crumbled petals. Calyx is with a 0.2 cm tube and 0.3 cm spread lobes. Fruit is a small brown colored round capsule. Fruit opens irregularly and splits into four sections at maturity and is many seeded. Seeds are about 3 mm across, numerous, smooth, pyramidal, hard and thick seed coat with brownish coloration. It is commonly called as Henna and belongs to the family Lythraceae (Chaudhary et al., 2012). The bark extract exhibited absolute toxicity against 13 ringworm fungi, the activity has remained the same even after autoclaving at high temperature and on long storage (Syamsudin and Winarno, 2008). MIC values ranging between 3.12–12.5 mg/mL demonstrates the natural antidermatophytic activity of the extracts (Sharma et al., 2011) and on comparison to other plants such as *Juglans regia*, *Pistacia lentiscus*; *L.inermis* appeared more active in inhibiting the dermatophytes with 18.87±0.58 mm of zone of inhibition (Mansour –Djaalab et al., 2012).

2.3.1.1.20 Mentha arvensis

Mentha arvensis L. commonly called mint and belongs to the family Lamiaceae. The essential oil of *M.arvensis* when assessed against the *T.rubrum* and *M.gypseum* exhibited strong activity and the formulation as ointment combining with essential oils from *Chenopodium ambrosioides*, *Cymbopogon citrates*, *Caesulia axillaris* and *Artemisia nela-grica* were able to cure experimental ringworm in guinea pigs within 7 to 12 days (Kishore et al., 1993). Bringing two or more oils or crude extracts together, sometimes may forbid or may enhance the activity, the activity is named either as synergism or as antagonism. *M.arvensis* show evidence of possessing synergistic effect with the above mentioned essential oils.

2.3.1.1.21 Moringa oleifera

Moringa oleifera Lam. are eaten and is cultivated for foods and medicinal purposes (Olson, 2002). It is commonly called as horse radish, benzolive, drumstick. The plant is a perennial soft wood native to the Sub-Himalayan tracts of India, Pakistan, Bangladesh, and Afghanistan. *Moringa oleifera* is a small, fast-growing evergreen or deciduous tree that usually grows up to 10 or 12 m in height. It has a spreading, open crown of drooping, fragile branches, feathery foliage of trip innate leaves, and thick, corky, whitish bark. The essential oil of *M.oleifera* exhibited prominent anti-dermatophytic activity (Chuang et al., 2007) where the GC-MS analysis was observed with the presence of 44 compounds. The leaf extracts were extracted with water, methanol and 70% ethanol, among them ethanolic extracts greatly minimized the growth of the tested dermatophytic organisms such as *M.ferrugineum*, *T.soudanensse*, *T.tonsurans*, *T.verrucosum* and *T.mentagrophytes* (Ayanbimpe et al., 2009; Oluduro, 2012).

2.3.1.1.22 *Murraya koenigii*

Murraya koenigii Spreng. belongs to the family Rutaceae, commonly known as curry-leaf tree, is a native of India, Sri Lanka and other south Asian countries. It is found almost everywhere in the Indian subcontinent, it shares aromatic nature, more or less deciduous shrub or tree up to 6 m in height and 15–40 cm in diameter with short trunk, thin smooth grey or brown bark and dense shady crown (Handral et al., 2012). The ethanolic extract exerted significant effect on the hyphal morphology, condition and germination of *T.mentagrophytes* and *M.gypseum*. The assessed total lipid and ergosterol content were found to have decremented compared to the normal level, hence could be believed that the crude extract had influence on both the lipid and sterol synthesis. It was also found to inhibit the lipase secretion in the tested organisms (Jayaprakash and Ebenezer, 2012). From ancient time, the curry leaves are employed as an ingredient in day-today food and these reports have proven the medicinal value of Indian foods.

2.3.1.1.23 *Occimum gratissimum*

Occimum gratissimum L. is an aromatic, perennial shrub belonging to the family Lamiaceae. It is commonly known as Scent leaf or Clove basil and is found in many tropical countries. It is 1–3 m tall; stem erect, round-quadrangular, much branched, glabrous or pubescent, woody at the base, often with epidermis peeling in strips. Silva et al. (2005) evaluated and reported the antifungal activity of the hexanic fraction and the pure compound eugenol against the *M.canis*, *M.gypseum*, *T.rubrum* and *T.mentagrophytes*. Growth was completely inhibited by the hexane extract at the concentration of 125 μg/mL, whereas the eugenol shows only 80% of inhibition. The potentiality of the leaves on the dermatophytic inhibition was also stated by Mbakwem-Aniebo et al. (2012).

2.3.1.1.24 Occimum sanctum

Occimum sanctum L. is commonly called Basil an annual herb belonging to the mint family has been cultivated for thousands of years and has become an essential ingredient in many cooking traditions. Basil is a member of the Lamiaceae, used both as a culinary and ornamental herb. The organic extracts of the plant leaves exhibited broad spectrum of inhibitory action against most of the dermatophyte species such as *T.rubrum*, *T.mentagrophytes*, *T.tonsurans*, *M.canis*, *E.floccosum*, *M.nanum* and *M.gypseum* (Das et al., 2010). MIC and MFC were calculated based on the NCCLS method and the particular fraction obtained from methanol extract showed precise activity against *T.mentagrophytes* with MIC value about 125±25 μg/mL concentration (Balakumar et al., 2011).

2.3.1.1.25 Phyllanthus amarus

Phyllanthus amarus Schum & Th. (Euphorbiaceae) is an annual, glabrous herb grows up to 15–60 cm high. Has an erect stem, naked below and slender and spreading leaf branches above. Leaves are numerous, subsessile, pale green, often distichously imbricating, glaucous below, elliptic to oblong, obtuse, and stipules subulate. Flowers arise in leaf axis, very numerous, males 1–3 and females solitary. Sepels of male orbicular and obovate to oblong in females. Stamens 3, anthers sessile and in a short column. Disc of male minute glands and of females annular and lobed. Capsules depressed globose, smooth and hardly 3 lobed. Seeds are 3-gonous, rounded and with longitudinal regular parallel ribs on the back. The chloroform extract of aerial parts showed maximum inhibition against the dermatophyte *M.gypseum* at 4000 ppm concentration and greatly reduced the sporulation of the organism (Agrawal et al., 2004).

2.3.1.1.26 Piper betle

Piper betle L. belongs to the family Piperaceae, is commonly used as a cultural symbolism and the leaves of this plant are economically and medicinally important. Water extracts showed MIC of 9.3 mg/mL against *T.rubrum* and *T.mentagrophytes* (Vaijayanthimala et al., 2004). Since the ethanolic extract of *P.betle* leaves showed promising activity against the zoonotic dermatophytes (*M.canis*, *M.gypseum* and *T.mentagrophyte*),

Trakranrungsie et al. (2006) has formulated the extract into cream (Pb cream). The Pb cream containing 80 μg of *P.betle* extract revealed comparable zones of inhibition with the ketoconazole. It was found that the bioactive compound hydroxychavicol from the chloroform fraction collected from aqueous extract exhibited anti-dermatophytic activity showing MIC value between 7.81 to 62.5 μg/mL (Ali et al., 2010). The chloroform extract showed 46 mm zone of inhibition against the *T.tonsurans* whereas the organism become more susceptible when treated with extract containing both chloroform extract of *P.betle* and *Allamanda cathertica* have increased the zone to 51 mm (Sharma et al., 2011).

2.3.1.1.27 *Piper longum*

Piper longum L. (Family: Piperaceae) grows all over India, in evergreen forests and is cultivated in Assam, Tamil Nadu and Andhra Pradesh. A small shrub with a large woody root and numerous creeping, jointed stems, thickened at the nodes. The leaves are alternate, spreading, without stipules and blade varying greatly in size. The lowest leaves are 5–7 cm long, whereas, the uppermost 2–3 cm long. The flowers are in solitary spikes. The fruits, berries, in fleshy spikes 2.5–3.5 cm long and 5 mm thick, oblong, blunt and blackish green in color. The mature spikes collected and dried, form the commercial form of pippali and the root radix is known as pippalimula. The chloroform extract of the leaf showed better activity when compared to the petroleum ether, methanol and water extracts against *T.rubrum*, *T.tonsurans*, *M.fulvum* and *M.gypseum* where the MIC was recorded at 5 mg/mL. The major compound showing the bioactivity was identified as 1,2-benzenedicarboxylic acid bis-(2-ethylhexyl) ester, 2,2-dimethoxybutane, and β-myrcene obtained through the analysis of GC-MS data from the fractions collected using silica gel column chromatography (Das et al., 2012).

2.3.1.1.28 *Pogostemon parviflorus*

Pogostemon parviflorus Benth. belongs to the family Lamiaceae, is a suffruticose shrubm 1.2–1.8 m high, stem and branches obtusely quadrangular, usually purple. Leaves 7.5–18 cm long, broadly ovate, acute or acuminate, coarsely and irregularly doubly-toothed, base cuneate. Flowers in dense pubescent spikes, forming pyramidal lax panicles. Corolla white, stamens exerted, filaments purple except just below the anthers, bearded with purple hairs. The ethanolic extract of *P.parviflorus* leaf completely prohibited the growth of *T.mentagrophytes*, *M.canis* and *M.gypseum* with minimum inhibitory concentration (MIC) values between 2.5–10 mg/mL (Sadeghi-Nejad and Deokule, 2010).

2.3.1.1.29 Psidium guajava

Psidium guajava is a medium sized tree with evergreen, opposite, aromatic short-petioled leaves. The inflorescence axillary 1–3 flowered trees are used for treatment of various disease conditions especially in the developing countries (Geidam et al., 2007). Hexane extracts of *P.guajava* leaves was found to inhibit *Trichophyton rubrum, Trichophyton tonsurans, Sporotrix schenckii, Microsporum canis* showing zone of inhibitions at or greater than 10 mm (Beatriz et al., 2012).

2.3.1.1.30 Punica granatum

Punica granatum L. (Punicaceae) is a small multi-stemmed shrub/tree 5–10 m tall. Canopy open, crown base low. Stem woody and spiny, bark smooth and dark grey. Leaves simple, 2–8 cm long, oblong or obovate, glabrous, oppositely placed, short-petioled surface shining. Flowers regular, solitary or in fascicles at apices, 4–6 cm. Petals lanceolate, 5–7, wrinkled and brilliant orange-red. Hypanthium colored, 58 lobed. Anthers numerous. Calyx persistent. Fruit a round berry, 5–12 cm, pericarp leathery. Interior compartmentalized with many pink-red sections of pulp-like tissue, each contains a seed grain. Fruits globose with persistent callipe and a coriaceous woody rind. Seeds numerous, angular with fleshy testa, 1.3 cm long (Arun and Singh, 2012). In 2008, Dutta et al. found that water extract of *P.granatum* was detrimental to dermatophytes and this was confirmed with result obtained by Shrivastav et al. (2013) against the growth on *T.tonsurans, T.mentagrophytes, T.rubrum, T.equinum, M.gypseum, M.nanum, M.audouinii*.

2.3.1.1.31 Rosmarinus officinalis

Rosmarinus officinalis L. belongs to the family Lamiaceae, is a small evergreen which grows wild in most Mediterranean countries, reaching a height of 1.5 m. The main producers are Italy, Dalmatia, Spain, Greece, Turkey, Egypt, France, Portugal and North Africa. Essential oils of *R.officinalis*, known as rosemary oils, are obtained by steam distillation of the fresh leaves and twigs, and the yields range from 0.5–1.0%. Color of the oil is almost colorless to yellow liquid with a pleasant odor and the major constituents are described as α-pinene, 1,8-cineole and camphor (Tiwari and Virmani, 1987). Rosemary oil exhibited strong antidermatophytic activity against the *M.audounii*, *T.rubrum*, *T.violaceum*, *T.tonsurans*, *T.verrucosum*, *T.mentagrophyte*, and *E.floccosum* at 1% concentration (Muyima and Nkata, 2005).

2.3.1.1.32 *Rubia cordifolia*

Rubia cordifolia L. is a climbing or scrambling herb, with red rhizomatous base and roots. The plant is commonly known as 'Indian Madder' and sold under the trade name 'manjistha'. Stem is quadrangular, divaricately branched, glabrous or prickly-hispid, especially on the angles. Leaves are 3.8–9 X 1.6–3.5 cm long, arranged in a whorl of four, cordate-ovate to ovate-lanceolate, 3–9 palmately veined, upper surface mostly glabrous and rough (Devi Priya and Siril, 2013). The petroleum ether extract of root exhibited antifungal activity against *T.rubrum* showing MIC at 0.253 mg/mL and further the extract was fractionized and analyzed through TLC, which showed the presence of anthraquinones (Gandhi, 2006).

2.3.1.1.33 *Solanum indicum*

Solanum indicum Linn. (Synonym: Solanum anguivi) belongs to the family Solanaceae commonly known as Byakur, is a bushy herb containing prickly spikes in the stem and available throughout the India and all over the tropical and subtropical regions of the world (Chopra et al., 1992). The leaves extracted with chloroform, methanol and water were tested against five dermatophytes such as *T.mentagrophytes*, *T.rubrum*, *T.tonsurans*, *M.gypseum* and *M.fulvum* and found high activity in the chloroform extract showing MIC value about 2.5 to 5 mg/mL, followed by the methanol extract, whereas the water extract exhibit in converse (Kotoky et al., 2012).

2.3.1.1.34 *Solanum nigrum*

Solanum nigrum L. (Solanaceae) is commonly called black nightshade, is an annual herbaceous plant, which can reach upto 100 cm in height. The stem may be smooth or bear small hairs. The flowers usually white in color, have five regular parts and are up to 0.8 cm wide. The leaves are alternate and somewhat ovate with irregularly toothed wavy margin and can reach 10 cm in length and 5 cm in width. The fruit is a round fleshy berry up to 2 cm in diameter and yellowish when ripe. The seeds are brown and numerous (Akubugwo et al., 2007). Ali-Shtayeh and Abu Ghdeib (1998) reported the anti-dermatophytic activity against *M.canis*, *T.mentagrophytes* and *T.violaceum* showing MIC value about 8.0 ± 2.82, 61.5 ± 10.13 and 81.2 ± 10.83 µg/mL respectively. Ethanolic and aqueous extract significantly inhibited the *Trichophyton* species and no effect on *Epidermophyton floccosum* (Shamin et al., 2004).

2.3.1.1.35 *Tamarindus indica*

Tamarindus indica L. of the Fabaceae, is an important food in the tropics. It is a frost-tender, tropical, evergreen tree, commonly called tamarind. The tree is densely foliated with pale green, compound, feathery leaflets which give the broad, spreading crown a light, airy effect. Tamarind may reach heights of 65 feet and a spread of 50 feet but is more often seen smaller. The delicate leaflets cast a diffuse, dappled shade which will allow enough sunlight to penetrate for a lawn to thrive beneath this upright, dome-shaped tree (Caluwe et al., 2010). *Eucalyptus globulus* and *Tamarindus indica* both produce allelochemicals which are already known for its antidermatophytic activity against *M.gypseum*, *T.terrestre* and *M.gypseum* (Sharma et al., 2008).

2.3.1.1.36 *Terminalia arjuna*

Terminalia arjuna belongs to the family Combretaceae is a tree with simple leaf, smooth and thick bark. Flowers are small, regular, sessile, cup-shaped, polygamous, white, creamy or greenish-white and robustly honey-scented. The inflorescence are short axillary spikes or small terminal panicles and fruits are obovoid-oblong, dark brown to reddish brown fibrous woody, indehiscent drupe (Khan et al., 2013). The bark of the plant extracted with acetone, 95% alcohol and methanol were evaluated against five dermatophyes such as *T.mentagorphytes*, *T.rubrum*, *T.tonsurans*, *M.gypseum* and *M.fulvum* using agar cup diffusion method. The methanol extract exhibited significant inhibition and showed MIC value about 75 μg/mL for *T.tonsurans*, 125 μg/mL for *T.mentagrophytes*, 250 μg/mL for *T.rubrum*, *M.fulvum* and 2000 μg/mL for *M.gypesum* (Bhattacharyya and Jha, 2011).

2.3.1.1.37 *Thymus vulgaris*

Thymus vulgaris L. (family: Labiatae or Lamiaceae) also known as common thyme, a plant native to the Mediterranean region has long been used as a source of the essential oil (thyme oil) and other constituents (e.g., thymol, flavanoid, caffeic acid and labiatic acid) derived from the different parts of the plant (Hudaib et al., 2002). The essential oil of *T.vulgaris* contains thymol as a major component and it was observed that the oil with antidermatophytic activity against *T.mentagrophytes* (Mota et al., 2012).

2.3.1.1.38 *Tridax procumbens*

Tridax procumbens Linn. belongs to the family Compositae. It is commonly known as 'Common button' or 'Coat button' and it is a weed found throughout India. A hispid, procumbent herb with woody base sometime rooting at the node, upto 60 cm high. Leaves are ovate-lanceolate 2 to 7 m and lamina pinnatisect, sometimes three lobed; flowers in small, long peduncled heads; achenes 1.5–2.5 mm long x 0.5 –1 mm in diameter and densely ascending pubescent; persistent ; bristles of disc achenes alternately longer and shorter, 3.5–6 mm in length (Kuldeep and Pathak, 2013). *T.procumbens* along with *Lantana camara* and *Capparis decidua* in the form of ointment were applied topically in the *T.mentagrophytes* infected animal model. The *in vitro* analysis showed the root and leaf extracts were effective against the pathogen showing MIC value at 0.312 mg/mL and 0.625 mg/mL (Bindu et al., 2011).

2.3.1.1.39 *Wrightia tinctoria*

Wrightia tinctoria R. Br. belongs to the family Apocynaceae is commonly called "indrajav" is a small and deciduous tree which grows up to 10 m with milky latex, scaly, smooth and ivory colored bark. Leaves are about 8–15 cm, opposite, variable, elliptic lanceolate or oblong lanceolate. Leaves are acute or rounded at the base, acuminate at the apex, petioles 5 mm long. Flowers are usually seen at the tip of branches with 6 cm long cymes, white with fragrance. It is widely distributed in India and Burma (Anusharaj et al., 2013). It was found that the leaves had potent antidermatophytic activity against most of the organisms such as *T.tonsurans*, *T.mentagrophytes* in addition to *T.rubrum* and *E.floccosum* showing IC_{50} value at 2 mg/mL (Kannan et al., 2006).

Ponnusamy et al. (2010) tested the hexane and chloroform extracts of leaves against *T.rubrum*, *E.floccosum* at 0.5 mg/mL concentration and the major compound indirubin also found to exhibit activity against *E.floccosum* (MIC: 6.25 μg/mL); *T.rubrum* and *T.tonsurans* (MIC: 25 μg/mL). Leaves also extracted with ethanol completely inhibit the pathogen *E.floccosum* at 500–100 ppm (Ranjani et al., 2012).

2.3.1.1.40 Zingiber officinale

Zingiber officinale Roscoe (Zingiberaceae) commonly called ginger, is a perennial rhizome which creeps and increases in size underground. Ginger is highly medicinal and used in the food preparations. Ginger is native to China and India. The essential oil of *Z.officinale* was found with strong antifungal activity against *T.rubrum* and *M.gypseum* showing the zone of inhibition about 72 mm and 69 mm and MIC value at 0.05 μg/mL and 0.06 μg/mL, respectively. Whereas the ginger oil in mixture with turmeric oil exhibited an excellent activity (82 mm, MIC 0.02 μg/mL against *T. rubrum* and 79 mm, MIC 0.04 μg/mL against *M. gypseum*) emphasizing its synergistic potentiality (Meenakshi and Richa, 2010).

2.4 CONCLUSION

Herbal remedies are considered the oldest forms of health care known to mankind on this earth. In India, it is reported that traditional healers use 2500 plant species in curing the several types of ailments and about 100 species of plants are regularly serve as sources of medicine (Shenq-Ji, 2001). Modern medicines basically depend on the traditional systems of medicine which is maintained and evolved over centuries among various tribal communities and are still maintained as a great traditional knowledge base in herbal medicines and retained by various indigenous groups around the world. Knowledge on the usage of herbal medicines is transferred from generation to generation. About 70 plant species belonging to 42 families are used for various purposes by tribal people of Western Ghats. This chapter is mainly focused on the flora of Western Ghats whose medicinal values were technically proven especially in treating the skin infection caused by a fungal group called Dermatophytes. In ICMR program a research group under the investigation of Dr. S.D. Kholkute surveyed and recorded the seasonal and non-seasonal plants along with the voucher specimens, botanical identity which resulted as 500 plants database. However, only 41 plants, i.e., 8.2% of the plants have been found to be scientifically proven toxic against the dermatophytes. Therefore it is concluded that more effort is required in preserving our nature's gift and its appropriate utilization in healing various kind of ailments. Futuristic aspect is primarily to bring out the scientific knowledge on the synergism exhibited by plant extracts and oils, secondly to evaluate the ayurvedic, siddha, unani and other plant based preparations for their bioactivity, thereby facilitating the improvement in the currently existing treatments.

KEYWORDS

- **alternative and complementary medicines**
- **antidermatophytic activity**
- **dermatophytes**
- **fungal skin infections**
- ***in vitro* analysis**
- **Western Ghats**

REFERENCES

Adamu, H. M., Abayeh, O. J., Ibok, N. U., Kafu, S. E. Antifungal activity of extracts of some Cassia, Detarium and Ziziphus species against dermatophytes Nat. Prod. Radiance 2006, 5(5), 357–360.

Agrawal, A., Srivasta, S., Srivastava, J. N., Srivastava, M. M., Inhibitory effect of the plant *Boerhavia diffusa* L. against the dermatophytic fungus *Microsporum fulvum*. J. Environ. Biol. 2004b, 25(3), 307–311.

Agrawal, A., Srivastava, S., Srivastava, J. N., Srivasava, M. M. Evaluation of inhibitory effect of the plant *Phyllanthus amarus* against dermatophytic fungi *Microsporum gypseum* Biomed. Environ. Sci. 2004, 17(3), 359–365.

Agrawal, A., Srivastava, S., Srivastava, M. M. Antifungal activity of *Boerhavia diffusa* against some dermatophytic species of Microsporum Hindustan Antibiot. Bull. 2004a, (1–4), 45–46.

Ajello, L., Georg, L. K., Kaplan, W., Kaufman, L. Laboratory manual for Medical Mycology 1966, US.

Akubugwo, I. E., Obasi, A. N., Ginika, S. C. Nutritional potential of the leaves and seeds of Black Nightshade-*Solanum nigrum* L. *Var virginicum* from Afikpo-Nigeria Pakistan J. Nutr. 2007, 6(4), 323–326.

Ali-Shtayeh, M. S., Abu Ghdeib, S. I. Antifungal activity of plant extracts against dermatophytes Mycoses 1998, 42, 665–672.

Ali, I., Khan, F. G., Suri, K. A., Gupta, B. D., Satti, N. K., Dutt, P., Afrin, F., Qazi, G. N., Khan, I. A. *In vitro* antifungal activity of hydroxychavicol isolated from *Piper betle* L Annals of Clinic. Microbiol. Antimicrob. 2010, 9(7), 1–9.

Aniebo, C., Okoyomo, E. P., Ogugbue, C. J., Okonko, I. O. Effects of *Jatropha curcas* leaves on common dermatophytes and causative agent of *Pityriasis versicolor* in rivers state, Nigeria Nature and Science. 2012, 10(12), 151.

Anonymous. Common fungal infections of the skin. Pharmawise. Spring 2006. 10(2), 1–6.

Anusharaj, Chandrashekar, R., Prabhakar, A., Rao, S. N., Santanusaha. *Wrightia tinctoria*: An overview. Journal of Drug Delivery and Therapeutics 2013, 3(2), 196–198.

Arun, N., Singh, D. P. *Punica granatum*: A review on pharmacological and therapeutic properties Int. J. of Pharm. Sci. and Res. 2012, 3(5), 1240–1245.

Ayanbimpe, G. M., Ojo, T. K., Afolabi, E., Opara, F., Orsaah, S., Ojerinde, O. S. Evaluation of extracts of *Jatropha cucas* and *Moringa oleifera* in culture media for selective inhibition of saprophytic fungal contaminants J. Clinic. Lab. Anal. 2009, 23, 161–164.

Bakhru, H. C. Herbs that heal, Orient Longman Ltd., 1997; p 17.

Balakumar, S., Rajan, S., Thirunalasundari, T., Jeeva, S. Antifungal activity of *Aegle marmelos* (L.) Correa (Rutaceae) leaf extract on dermatophytes Asian Pac. J. Trop. Biomed. 2011, 1(4), 309–312.

Balakumar, S., Rajan, S., Thirunalasundari, T., Jeeva, S. Antifungal activity of *Ocimum sanctum* Linn. (Lamiaceae) on clinically isolated dermatophytic fungi Asian Pacific J. Trop. Med. 2011, 654–657.

Balakumbahan, R., Rajamani, K., Kumanan, K. *Acorus calamus*: An overview J. Med. Plants Res. 2010, 4(25), 2740–2745.

Beatriz, P. M., Ezequie, V. V., Azucena, O. C., Pilar, C. R. Antifungal activity of *Psidium guajava* organic extracts against dermatophytic fungi J. Med. Plants Res. 2012, 6(41), 5435–5438.

Bhattacharyya, P. N., Jha, D. K. Antidermatophytic and antioxidant activity of *Terminalia arjuna* (roxb.) Wight & Arn. Bark Int. J. Pharm. and Biol. Archives 2011, 2(3), 973–979.

Bindu, S., Padma, K., Suresh, C. J. Topical treatment of dermatophytic lesion on mice (Mus musculus) model Indian J. Microbiol. 2011, 51(2), 217–222.

Caluwe, E., Halamova, K., Damme, P. V. *Tamarindus indica* L.- A review of traditional uses, phytochemistry and pharmacology Arika Focus 2010, 23(1), 53–83.

Chaudhary, G., Goyal, S., Poonia, P. *Lawsonia inermis* Linnaeus: A Phytopharmacological review Int. J. Pharm. Sci. Drug Res. 2010, 2(2), 91–98.

Chopra, R. N., Nayer, S. L., Chopra, I. C. Glossary of Indian Medicinal Plants. New Delhi: PID, 1992, CSIR.

Chuang, P. H., Lee, C. W., Chou, J. Y., Murugan, M., Shieh, B. J., Chen, H. M. Antifungal activity of crude extracts and essential oil of *Moringa oleifera* Lam. Bioresource Technology 2007, 98, 232–236.

Chudiwal, A. K., Jain D. P., Somani R. S. *Alpinia galanga* Willd. – An overview on phytopharmacological properties Ind. J. Nat. Prod. Res. 2012, 1(2), 143–149.

Current, W. L., Tang, J., Boylan, C., Watson, P., Zeckner, D., Turner, W., Rodriguez, M., Ma, D., Radding, J. Glucan biosynthesis as a target for antifungals; the echinocandin class of antifungal agents. In The discovery and mode of action of antifungal drugs; Dixon, G. K., Ed., L. G. Copping and D. W. Hollowmon, BIOS Scientific Publishers, Ltd., Oxford. 1995; p 143–160.

Das, J., Buragohain, B., Srivastava, R. B. *In vitro* evaluation of *Ocimum sanctum* leaf extract against dermatophytes and opportunistic fungi Asian J. Microbiol. Biotech. & Env. Sci. 2010, 12(4), 789–792.

Das, J., Jha, D. K., Policegoudra, R. S., Mazumder, A. H., Das, M., Chattopadhyay, P., Singh, L. Isolation and characterization of antidermatophytic bioactive molecules from *Piper longum* L. leaves Indian J. Microbiol. 2012, 52(4), 624–629.

Devi Priya, M., Siril, E. A. Pharmacognostic studies on Indian Madder (*Rubia cordifolia* L.). J. Pharmacognosy Phytochem. 2013, 1(5), 112–119.

Duraipandiyan, V., Al-Harbi, N. A., Ignacimuthu, S., Muthukumar, C. Antimicrobial activity of sesquiterpene lactones isolated from traditional medicinal plant, *Costus speciosus* (Koen ex.Retz.) sm. BMC Complement. and Alter. Med. 2012, 12(13): 1–6.

Duraipandiyan, V., Ignacimuthu, S. Antifungal activity of traditional medicinal plants from Tamil Nadu, India Asian Pacific J. Trop. Biomed. 2011, S204–S215.

Dutta, B. K., Rahman, I., Das, T. K. Antifungal activity of Indian plant extracts: Antimyzetische Aktivitat indischer Pflanzenextrakte Mycoses 2008, 41(11–12), 535–536.

Ferdowsi, H., Afshar, S., Rezakhani, A. A comparison between the routine treatment of equine dermatophytosis and treatment with Garlic-Aloe vera gel Int. Res. J. Appli. and Basic Sci. 2012, 3(11), 2258–2261.

Gallis, H. A., Drew, R. H., Pickard, W. W. Amphotericin B: 30 years of clinical experience Ref. Infect. Dis. 1990, 12, 308–329.

Gandhi, A. S. Antifungal activity of *Rubia cordifolia* Pharmacology drug profiles 2006, 1–6.

Garcia, V. M. N., Gonzalez, A., Fuentes, M., Aviles, M., Rios, M. Y., Zepeda, G., Rojas, M. G. Antifungal activities of nine traditional Mexican medicinal plants J. of Ethnopharm. 2003, 87, 85–88.

Geidam, Y. A., Ambali, A. G., Onyeyili, P. A. Preliminary phytochemical and antibacterial evaluation of crude aqueous extract of *Psidium guajava* leaf J. Appl. Sci. 2007, 7, 511–514.

Ghannoum, M. A., Rice, L. B. Antifungal agents: mode of action, mechanisms of resistance, and correlation of these mechanisms with bacterial resistance Clinic. Microbiol. Rev. 1999, 12(4), 501–517.

Gunalan, G., Saraswathy, A., Krishnamurthy, V. Antimicrobial activity of medicinal plant *Bauhinia variegate* Linn. Int. J. Pharm. Biol. Sci. 2011, 1(4), 400–408.

Handral, H. K., Pandith, A., Shruthi, S. D. A review on *Murraya koenigii*: Multipotent medicinal plant Asian J. Pharm. Clinic. Res. 2012, 5(4), 5–14.

Hashmat, I., Azad, H., Ahmed, A. Neem (*Azadirachta indica* A. Juss) – A nature's drugstore: An overview Int. Res. J. Biol. Sci. 2012, 1(6), 76–79.

http://www.astrogle.com/ayurveda/curing-diseases-with-achyranthes-aspera-prickly-flower.html (assessed date- 20.11.2013

Hudaib, M., Speroni, E., Pietra, A. M., Cavrini, V. GC/MS evaluation of thyme (*Thymus vulgaris* L.) oil composition and variations during the vegetative cycle. J. Pharm. Biomed. Anal. 2002, 29, 691–700.

Jain, S., Patil, U. K. Phytochemical and pharmacological profile of *Cassia tora* Linn. – An Overview Ind. J. Nat. Prod. Resources, 2010, 1(4), 430–437.

Janssen, A. M., Scheffer, J. J. Acetoxychavicol acetate, an antifungal component of *Alpinia galanga* Planta Med. 1985, 51(6), 507–511.

Jayaprakash, A., Ebenezer, P. Antifungal activity of curry leaf (*Murraya koenigii*) extract and an imidazole fungicide on two dermatophyte taxa J. Acad. Indus. Res. 2012, 1(3), 124–126.

Judith Stevens, Fungal skin infections. Continuum of Care. (assessed date-25.7.2013)

Kannan, P., Shanmugavadivu, B., Petchiammal, C., Hopper, W. *In vitro* antimicrobial activity against skin microorganisms. Acta Botanica Hungarica 2006, 48(3–4), 323–329.

Khan, Z. M. H., Faruquee, H., Shaik, M. Phytochemistry and pharmacological potential of *Terminalia arjuna* L. Medicinal Plant Res. 2013, 3(10), 70–77.

Kotoky, J., Sharma, K. K., Kalita, J. C., Barthakur, R. Antidermatophytic activity of *Solanum indicum* L. from North East India J. Pharm. Res. 2012, 5(1), 265–267.

Kovacs, S. O., Hruze, L. I. Superficial fungal infections. Postgraduate Medicine, 1995, 98(6), 61–75.

Krishnan, P. N., Decruse, S. W., Radha, R. K. Conservation of medicinal plants of Western Ghats, India and its sustainable utilization through *in vitro* technology *In vitro* Cell Dev. Biol. Plant 2011, 47, 110–122.

Kuldeep, G., Pathak, A. K. Pharmacognostic and phytochemical evaluation of *Tridax procumbens* Linn. J. Pharmacognosy. Phytochem. 2013, 1(5), 42–46.

Londonkar, R., Chinnappa Reddy, V., Abhay Kumar, K. Potential antibacterial and antifungal activity of *Achyranthes aspera* L. Recent Res. Sci. and Tech. 2011, 3(4), 53–57.

Maity, P., Hansda, D., Bandyopadhyay, U., Mishrra, D. K. Biological activities of crude extracts and chemical constituents of bael, *Aegle marmelos* (L.) Corr. Ind. J. Exp. Biol. 2009, 47, 849–861.

Mansour-Djaalab, H., Kahlouche-Riachi, F., Djerrou, Z., Serakta-Delmi, S., Hamimed, S., Trifa, W., Djaalab, I., Hamdipacha, Y. *In vitro* evaluation of antifungal effects of *Lawsonia inermis*, *Psitacia lentiscus* and *Juglans regia* Int. J. Med. Arom. Plants 2012, 2(2), 263–268.

Mansuang, W., Sompop, P., Yuvadee, W. Antifungal activity and local toxicity study of *Alangium salvifolium* subsp *hexaptalum*. Southeast Asian J Trop Med Public Health 2002, 33(3), 152–154.

Maslin, B. R. Introduction to Acacia. In Flora of Australia; Orchard, A. E., Wilson, A. J. G., Eds., ABRS/CSIRO Publishing: Melbourne, Australia, 2001, Vol. 11A, pp. 3–13.

Mbakwem-Aniebo, C., Onianwa, O., Okonko, I. O. Effects of *Ocimum gratissimum* leaves on common dermatophytes and causative agent of *Pityriasis versicolor* in rivers state, Nigeria J. Microbiol. Res. 2012, 2(4), 108–113.

Meenakshi, S., Richa, S. Synergistic Antifungal Activity of *Curcuma longa* (Turmeric) and *Zingiber officinale* (Ginger) Essential Oils Against Dermatophyte Infections J. of Essential Oil Bearing Plants 2011, 14(1), 38–47.

Mishra, R. P., Siddique, L. Antifungal properties of *Anthocephalus cadamba* fruits Asian J. Plant Sci. Res. 2011, 1(2), 81–87.

Mosunmola, O. J., RamotaRemi, R. A., Abayomi, B. T., Olaosebikan, M. S., Charles, N., Afolabi, O. Susceptibility of dermatophytes to *Aloe vera* juices using agar diffusion and broth dilution techniques Amer. J. Res. Comm. 2013, 1(8), 53–62.

Mota, K. S. L., Pereira, F. O., Oliveira, W. A., Lima, I. O., Lima, E. O. Antifungal activity of Thymus vulgaris L. Essential oil and its constituent phytochemicals against Rhizopus oryzae: Interaction with ergosterol Molecules, 2012, 17, 14414–14433.

Muyima, N. Y. O., Nkata, L. Inhibition of the growth of dermatophyte fungi and yeast associated with dandruff and related scalp inflammatory conditions by the essential oils of *Artemisia afra*, *Pteronia incana*, *Lavandula officinalis* and *Rosmarinus officinalis* J. Essential Oil-Bearing Plants 2005, 8(3), 224–232.

Niranjan, A., Tewari, S. K., Lehri, A. Biological activities of Kalmegh (*Andrographis paniculata* Nees) and its active principles- A review Indian J. Nat. Prod. Res. 2010, 1(2), 125–135.

Noble, S. L., Forbes, R. C., Diagnosis and Management of Common Tinea Infections. Kansas City, Missouri: American Academy of Family Physicians. 1998.

Olufunmiso, O., Olajuyigbe, Anthony, J., Afolayan. Pharmacological assessment of the medicinal potential of Acacia mearnsii De Wild.: Antimicrobial and toxicity activities Int. J. Mol. Sci. 2012, 13, 4255–4267.

Palacio, A. D., Garau, M., Gonzalez-Escalada, A., Calvo, M. T. Trends in the treatment of dermatophytosis Revista Iberoamericana de Micologia 2000, 148–158.

Ponnusamy, K., Chelladura, P., Ramasamy, M., Waheeta, H. *In vitro* antifungal activity of indirubin isolated from a South Indian ethnomedicinal plant *Wrightia tinctoria* R. Br. J. Ethnopharmacol. 2010, 132, 349–354.

Radhika, S. M., Michael, A. Antidermatophytic activity of Azadirachta indica and Acalypha indica Leaves: An *in vitro* study Int. J. Pharm. Bio. Sci. 2013, 4(4), 618–622.

Rajeswari, R., Umadevi, M., Sharmila Rahale, C., Pushpa, R., Selvavenkadesh, S., Sampath Kumar, K. P., Bhowmik, D. *Aloe vera*: The miracle plant its medicinal and traditional uses in India J. Pharmacog. Phytochem. 2012, 1(4), 118–124.

Ramya. R., Lakshmi Devi, N. Antibacterial, antifungal and antioxidant activities of *Andrographis paniculata* Nees. leaves Int. J. Pharmaceut. Sci. Res. 2011, 2(8), 2091–2099.

Rani, S., Sulakshana, G., Patnaik, S. *Costus speciosus*, an antidiabetic plant—Review J. Pharm. Res. 2012, 1(3), 52–53.

Ranjani, M., Deepa, S., Kalaivani, K., Sheela, P. Antibacterial and antifungal screening of ethanol leaf extract of *Wrightia tinctoria* against some pathogenic microorganisms. Drug Invention Today 2012, 4(5), 365–367.

Rath, S., Mohanty, R. C. Antifungal screening of *Curcuma longa* and *Cassia tora* on dermatophytes Int. J. Life Sci. Biotech. and Pharm. Res. 2013, 2(4), 88–94.

Roa, R. R., Sagar, K. Invasive alien weeds of the Western Ghats: Taxonomy and Distribution. In Invasive alien plants an ecological appraisal for the Indian subcontinent; Bhatt, J. R., Singh, J. S., Singh, S. P., Tripathi, R. S., Kohli, R. K., CABI publishers, 2012; Chapter-12.

Sadeghi-Nejad, B., Deokule, S. S. Antidermatophytic activity of *Pogostemma parviflorus* Benth. Iranian J. Pharm. Res. 2010, 9(3), 279–285.

Sahu, A. N., Damiki, L., Nilanjan, G., Dubey, S. Phytopharmacological review of *Boerhaavia diffusa* Linn. (Punarnava) Phcog. Rev. 2008, 2(4), 14–22.

Santosh Kumar, S., Anil, S., Priyanka, S., Agrawal, R. D. Antidermatophytic activities of different plant parts extract against *Trichophyton rubrum* and *Candida albicans* dermatophytes isolated from HIV+ Ves of Jaipur District Rajasthan Asian J. Biochem. Pharm. Res. 2012, 1(2), 146–152.

Shamin, S., Ahmed, S. W., Azhar, I. Antifungal activity of Allium, Aloe and Solanum species Pharm. Biol. 2004, 42(7). 491–498.

Sharma, K. K., Saikia, R., Kotoky, J., Kalita, J. C., Das, J. Evaluation of antidermatophytic activity of *Piper betle*, *Allamanda cathertica* and their combination: an *in vitro* and *in vivo* study Int. J. of PharmTech Res. 2011, 3(2) 644–651.

Sharma, K. K., Saikia, R., Kotoky, J., Kalita, J. C., Devi, R. Antifungal activity of *Solanum melongena* L, *Lawsonia inermis* L. and *Justicia gendarussa* B. against dermatophytes Int. J. Pharm Tech. Res. 2011, 3(3), 1635–1640.

Sharma, P. C., Bhatia, V., Bansal, N., Sharma, A. A review on bael tree Nat. Prod. Radiance 2007, 6, 171–178.

Sharma, R., Upadhyaya, S., Singh, B. S., Singh, B. G. Inhibitory effect of allelochemicals produced by medicinal plants on dermatophytes. In Phytochemicals: a therapeutant for critical disease management; Khanna, D. R., Chopra, A. K., Prasad, G., Malik, D. S., Bhutiani, R. 2008, pp. 349–352.

Shenq-Ji, P. Ethnobotanical approaches of traditional medicine studies some experiences from Asia Pharam. Biol. 2001, 39, 74–79.

Shrivastav, V. K., Shukla, D., Parashar, D., Shrivastav, A. Dermatophytes and related keratinophilic fungi isolated from the soil in Gwalior region of India and *in vitro* evaluation of antifungal activity of the selected plant extracts against these fungi J. Med. Plants Res. 2013, 7(28), 2136–2139.

Silva, M. R., Oliveira, J. G., Femandes, O. F., Passos, X. S., Costa, C. R., Souza, L. K., Lemos, J. A., Paula, J. R. Antifungal activity of *Ocimum gratissimum* towards dermatophytes Mycoses 2005, 48(3), 172–175.

Srivastav, S., Pradeep Singh, Mishra, G., Jha, K. K., Khosa, R. L. *Achyranthes aspera*-An important medicinal plant: A review J. Nat. Prod. Plant Resour. 2011, 1(1), 1–14.

Subha, T. S., Gnanamani, A. Topical therapy of 1–2, 4, 5 trimethoxy phenyl 1' methoxypropionaldehyde in experimental *Tinea pedis* in Wistar rats Biol. Med. 2011, 3(2), 81–85.

Syamsudin, I., Winarno, H. The effects of Inai (Lawsonia *inermis*) leave extract on blood sugar level: An Experimental Study Res. J. Pharmacol. 2008, 2(2), 20–23.

Tariqo, S., Javed, A. Vitamin C content of Indian medicinal plants: A review literature Ind. Drugs, 1985, 23(2), 72–75.

Tewari, R., Virmani, O. P. Chemistry of Rosemary Oil: A Review. Central Institute of Medicinal and Aromatic Plants 1987, 9, 185–197.

Trakranrungsie, N., Chatchawanchonteera, A., Khunkitti, W. Antidermatophytic activity of *Piper betle* cream Thai. J. Pharmacol. 2006, 28(3), 16–20.

Trakranrungsie, N., Chatchawanchonteera, A., Khunkitti, W. Ethnoveterinary study for antidermatophytic activity of *Piper betle*, *Alpinia galanga* and *Allium ascalonicum* extracts *in vitro* Res. in Vet. Sci. 2008, 84, 80–84.

Vaijayanthimala, J., Rajendra Prasad, N., Anandi, C., Pugalendi, K. V. Anti-dermatophytic activity of some Indian med. plants 2004, 4, 1 26–31.

Vermes, A., Guchelaar, H. J., Dankert, J. Flucytosine: a review of its pharmacology, clinical indications, pharmacokinetics, toxicity and drug interactions J. Antimicrob. Chemotherapy 2000, 46, 171–179.

Vijayalakshmi, A., Tripathi, R., Ravichandiran, V. Characterization and evaluation of anti-dermatophytic activity of the essential oil from *Artemisia nilagirica* leaves growing wild in Nilgiris Int. J. Pharm. and Pharm. Sci. 2010, 2(4), 93–97.

Walsh, N. G., Entwisle, T. J. Flora of Victoria; Inkata Press: Melbourne, Australia, 1996; Vol. 3.

Weitzman, I., Summerbell, R. C. The Dermatophytes Clinic. Microbiol. Rev. 1995, 8(2), 240–259.

CHAPTER 3

IN VIVO SUPPRESSION OF SOLID EHRLICH CANCER VIA AG AND CO/AG MEDIATED PTT

IMAN E. O. GOMAA, MONA B. MOHAMED, and TAREK A. EL-TAYEB

CONTENTS

ABSTRACT

The use of nanoparticles (NPs) in cancer treatment has drawn considerable attention in different scientific fields. This is due to their altered physical, chemical and biological properties from their macrostructures. This study is directed to investigate the effect of silver "Ag" nanoparticles and cobalt core – silver shell "Co/Ag" nanoparticles on solid tumor ablation in albino mouse model by photothermal therapy.

Intra-tumoral injection of either Ag or Co/Ag NPs was done in 1 mm^3 tumor volume of Ehrlich tumor albino mouse model. Both treated and control mice were exposed to light emitting diode 460 nm wavelength and 250 mW power at 24 h post-NPs injection under general anesthesia. Several physical and biochemical parameters, as well as histopathological examinations, were conducted in order to verify the efficacy as well as safety levels of the applied photothermal strategy.

Unlike the tumor control animals, dark control (NPs treated mice without light exposure), and light control (NPs untreated mice with light exposure) groups, results of animals treated with AgNPs or Co/AgNPs mediated photothermal effect showed significant reduction in tumor volume without tumor recurrence for a follow-up period of six months. Co/AgNPs have less dark toxicity, but higher light toxicity than the AgNPs. Accordingly, Co/AgNPs is less toxic to tissue that is not exposed to light. There were no significant effect of AgNPs and Co/AgNPs on the physical and biochemical parameters, as well as histopathological aspects of albino mice.

The significance of tumor ablation, physical and biochemical parameters, as well as histopathological examinations may support the AgNPs and Co/AgNPs as good candidates for non-surgical method of solid tumor ablation.

3.1 INTRODUCTION

The conventional treatment modalities of solid tumors include surgery, chemotherapy and radiotherapy. In spite of being highly efficient at cancer treatment, the conventional modalities for cancer therapy have several drawbacks such as invasiveness for surgical therapy, lack of specificity for chemotherapy, and occasional limited sensitivity for radiotherapy (Gabriel,

2002). Application of hyperthermia effect using metallic nanoparticles is a promising approach for the control of solid tumors.

Hyperthermia is a state where cells absorb more heat than they can dissipate, which can be lethal to the cells. It has been found as a promising approach to cancer therapy because it directly kills cancer cells and indirectly activates anti-cancer immunity (van der Zee, 2002). Since focusing the heat on an intended region without damaging the healthy tissue is a critical issue that arose by the treatment with hyperthermia, targeting a specific region has been important. If nanoparticles are introduced into malignant tissue, hyperthermia can be directed to this specific tissue. Unlike the conventional hyperthermic techniques, some studies showed that using nanoparticles for hyperthermia helped cancer cells to reach the targeted temperature without damaging the surrounding tissue (Kikumori et al., 2009).

Nanoparticles encounter great possibilities in biomedicine and bioanalysis. Owing to the advantage of their attractive size (10 to 100 nm), it places them at dimensions smaller than or comparable to those of biological macromolecules, such as enzymes and receptors (Praetorius, 2007). As a result, nanoscale particles with diameters smaller than 50 nm can easily pass through most cells, while those smaller than 20 nm in diameter can transit out through blood vessels (McNeil, 2005). Additionally, from the great features making nanoparticles useful for bio-applications; they have low charge and hydrophilic surface, which are not recognized by the mononuclear phagocytic system and accordingly, have long half-life in the blood circulation (Asane, 2007).

Due to their special physicochemical properties, metal nanoparticles showed great progress in bioanalytical and medical applications, such as multiplexed bioassays, biomedicine, ultrasensitive biodetection (Nam et al., 2003) and bioimaging (Osaka, Tetsuya, et al., 2008). They have several biological applications including; drug delivery, magnetic resonance imaging "MRI" enhancement (Osaka, Tetsuya, et al., 2008), as well as applications in cancer treatment (Ruddon, 2007).

Silver nanoparticles are among the noble metallic nanomaterials that have received considerable attention due to their attractive physicochemical properties. The surface plasmon resonance and the large effective scattering cross-section of individual nanoparticles make them ideal candidates for biomedical applications (Liau, et al., 1997). Nowadays many studies focused on nontoxic silver nanoparticles synthesis for biological application (Schneidewind et al., 2012; Haberl et al., 2013). On the other hand, silver nano-shells of 40–50 nm outer diameter and 20–30 nm inner diameter using cobalt nanoparticles as sacrificial templates were also synthesized. As a result, the thermal reaction deriving force resulted from a large reduction potential gap between the Ag^{+}/Ag and the Co^{+2}/Co redox couples, which leads to the consumption of Co cores and the formation of a hollow cavity of Ag nano-shells. The UV spectrum of such nanostructure exhibits a distinct difference from that of solid nanoparticles, making it a good candidate for application in photo-thermal materials (Chen and Lian Gao, 2006).

Few studies have focused on using silver nanoparticles in tumor hyperthermia, and no study has been done for cobalt silver core shell nanoparticle system on solid tumor treatment. This study is focused on studying the efficiency of silver nanoparticles in naked form, and comparing it with that containing a cobalt core on solid tumor ablation, biochemical and histopathological aspects in albino mouse model.

3.2 MATERIALS AND METHODS

3.2.1 SYNTHESIS OF AGNPS AND CO/AGNPS

The Ag and Co/Ag nanoparticles were synthesized and characterized in the laboratories of the National Institute of Laser Enhanced Sciences (NILES). Their absorption wavelengths were detected by spectrophotometric analysis, and their size and shape were determined by transmission electron microscope before application on solid tumor hyperthermia.

3.2.2 IN VIVO STUDY

In vivo experiments were performed using Ehrlich ascites tumor cells. These are transplantable, poorly differentiated malignant tumor cells that appeared originally as a spontaneous breast adenocarcinoma in albino mice. Ehrlich tumor cells are grown in vivo at the abdominal region of albino mouse by intraperitoneal injection, forming ascetic cells, which form solid tumor mass upon subcutaneous injection.

3.2.2.1 ANIMALS STUDY GROUPS

In this study, normal laboratory albino mice at the age of 8 weeks and 20–25 gm body weight were obtained from the animal house of the National Cancer Institute (NCI) in Egypt. All animal experiments were performed following the regulations of the Research Ethics Committee at the German University in Cairo. A total number of 35 female mice "n=35" were used and divided into seven groups, each containing 5 mice. The first group included normal control mice that were not injected with any tumor cells. The second group represented the tumor control group containing mice subcutaneously transplanted with Ehrlich tumor cells at the right posterior thigh, but not treated with any nanoparticles or exposed to light. The third group was a light control group of mice having tumors that were exposed for 1 h to monochromatic light-emitting diode (LED) of wavelength 460 nm and power 250 mW. The fourth and sixth groups were dark control mice, which were intratumorally injected with Ag and Co/Ag nanoparticles respectively, but without being exposed to LED. Finally, the fifth and seventh groups contained treated animals at which tumors that were pre-injected with either Ag or Co/Ag particles were also locally exposed to the above indicated monochromatic blue LED.

Ehrlich tumor cells that have been intraperitoneally grown in albino mouse were extracted from the abdomen using 20 cc syringe (Figure 3.1A). A number of 1×10^6 tumor cells were suspended in 100 μl PBS, and subcutaneously injected in the thigh of albino mice forming solid tumor mass within about two weeks from onset of injection.

The nanoparticles suspension was centrifuged at 20000 rpm for 1 h, and the pellet was suspended in phosphate buffered saline (PBS) at a concentration of 0.4 gm/mL. Mice representing the dark control and treated groups were injected with 100 μL of either Ag or Co/Ag nanoparticles suspension in 1 cm^3 tumor mass size (Figure 3.1B).

At 24 h post-nanoparticles injection, mice were intraperitoneally injected with 60 mg/kg sodium thiopental (EIPICO) for general anesthesia. After 15 min, direct exposure of the solid tumor to LED (Photon LED DH48S) monochromatic blue light of 460 nm and 250 mW was applied for 1 h (Figure 3.1C). During light exposure, the general body temperature of each mouse was monitored every 10 min by measuring rectal temperature using a mercury thermometer (Figure 3.1D). Body temperature for normal mice with and without anesthesia was also monitored as a control.

For the next 30 days following light exposure, the perpendicular tumor dimensions were measured on daily basis for a week, then three times for two weeks, and twice a week for the rest of the follow-up period. The tumor regression volume was calculated using the modified ellipsoidal formula; [volume = (Length × Width^2 × π)/6] (Gümüşhan et al., 2013). Animals were sacrificed for ethical reasons when any of the tumor dimensions reached 2 cm length.

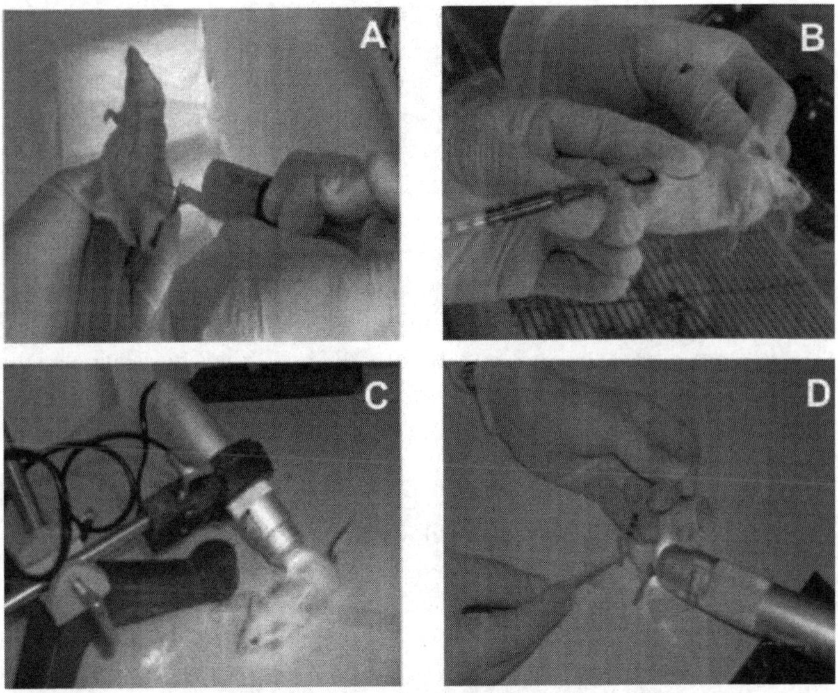

FIGURE 3.1 Extraction of Ehrlich ascites tumor cells from the abdominal region of a mouse having solid Ehrlich tumor (A). Intra-tumoral injection of nanoparticles (B). Monochromatic blue light exposure of the solid tumor induced at the right posterior leg of an anaesthetized albino mouse using LED of 460 nm and 250 mW (C). Monitoring of the rectal temperature of the mouse while being exposed to monochromatic blue light (D).

3.2.2.2 BIOCHEMICAL ANALYSIS

After 30 days follow-up of tumor volume, all mice were anaesthetized using diethyl ether before being dissected. Blood samples were collected by cardiac puncture using 3 cc syringes. Biochemical analyses including Alanine-Amino transferase and direct bilirubin were tested using (ALT, Diamond Diagnostics) and (Bl, Diamond Diagnostics) standardized kits respectively.

3.2.2.3 HISTOPATHOLOGY

Tumor biopsies were isolated from sacrificed animals, and fixed in 10% buffered formaldehyde for 24 h. Dehydration was then applied in ascending series of ethyl alcohol. Specimens were cleared in xylene then embedded in paraffin at 56°C in a hot air oven for 24 h. Tissue sections of 4 μm thick were prepared by a 'Slidge' microtome on glass microscopic slides. Tissue sections were deparaffinized and stained with hematoxylin and eosin for investigation of histopathological features.

3.2.3 STATISTICAL ANALYSIS

Statistical analysis was applied using the GraphPad Prism 5.0 software. Results were collected from 5 mice per animal group. Data was considered significant at P value < 0.05.

3.3 RESULTS

3.3.1 IN VIVO TUMOR ABLATION

Animals in the tumor control group had their tumors growing faster than those at any other group. Animals in the light control, as well as in the AgNPs or Co/AgNPs dark control groups, showed continuous but slower rate of tumor growth compared to those belonging to the tumor control group. Those in the treated groups showed continuous reduction in their sizes till complete disappearance (Figure 3.2).

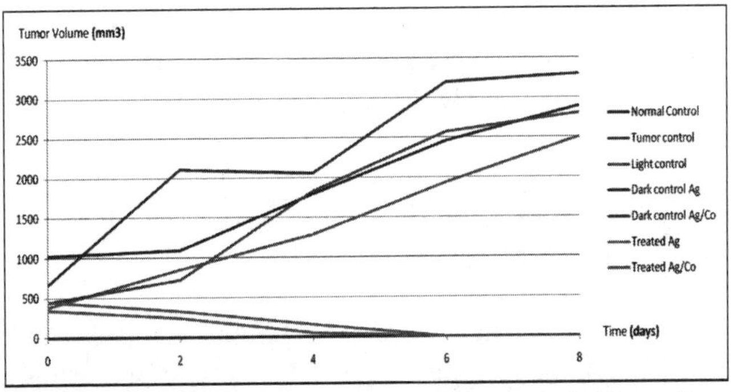

FIGURE 3.2 Representation of tumor growth delay by Ag or Ag/Co mediated PTT.

In the AgNPs and Co/AgNPs mediated photothermal effect groups of animals, tumors exposed to LED showed significant reduction in tumor volume. They acquired a black color right after 60 min of 460 nm and 250 mW monochromatic blue light exposure, then they turned flat and hard on day 4, and finally began to form a deep curvature at the tumor site on day 8. There has been no obvious difference between the treatment levels in the AgNPs or Co/AgNPs treated tumors (Figure 3.3 A–D).

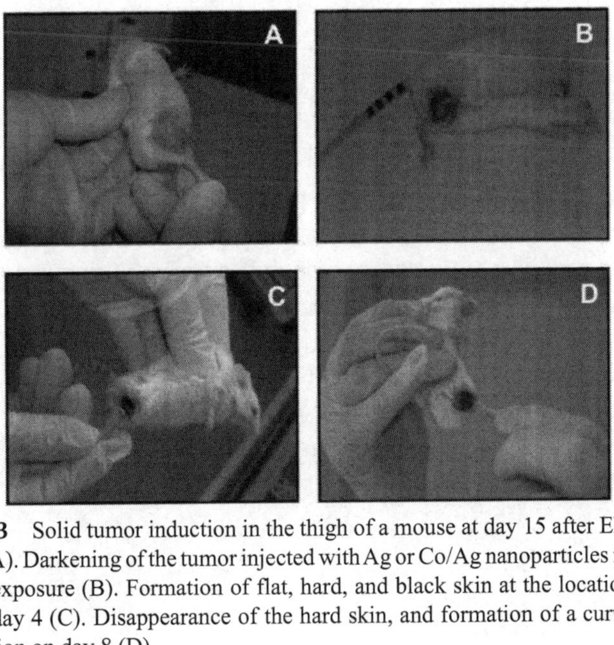

FIGURE 3.3 Solid tumor induction in the thigh of a mouse at day 15 after Ehrlich tumor induction (A). Darkening of the tumor injected with Ag or Co/Ag nanoparticles immediately after LED exposure (B). Formation of flat, hard, and black skin at the location of treated tumors on day 4 (C). Disappearance of the hard skin, and formation of a curvature at the tumor location on day 8 (D).

3.3.2 GENERAL BODY TEMPERATURE

The average variation in body temperature at different animal groups at 10 min. intervals for 1 h duration of light exposure has been investigated. The normal control group includes the average body temperature of both the normal control and tumor control mice with neither anesthesia nor exposure to LED. The anesthetized control group of animals involved the average body temperature of both the normal control as well as the tumor control mice under general anesthesia with sodium thiopental, but without being exposed to LED. On the other hand, the treated animals group showed the average body temperature of the AgNPs treated and Co/AgNPs treated mice under general anesthesia for 1 h during light exposure with 10 min intervals.

Compared to the 37.5°C, normal body temperature, the anesthetized control mice showed gradual decrease in general body temperature from 37.5°C to 35.5°C after 15 min post injection, then decreased to 34.5°C and finally reached stability at 34°C throughout the 1 h LED exposure. During light exposure of the treated mice, the body temperature slightly rose from 35.5°C at the beginning of exposure, and got stable around 36.5°C (Figure 3.4).

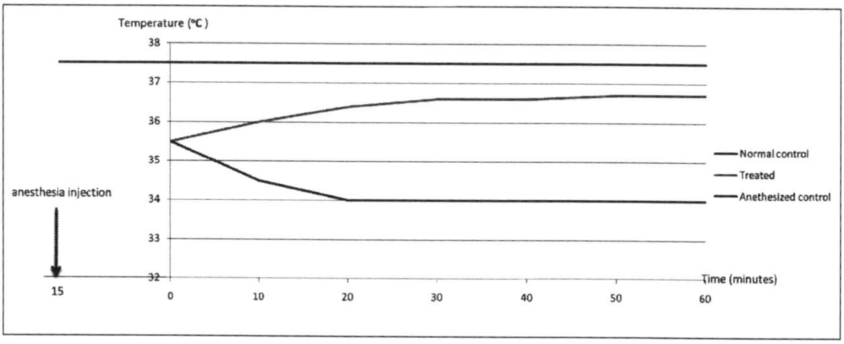

FIGURE 3.4 Graphical representation of the general body temperature of the treated, anesthetized control and normal control mice groups.

3.3.3 BIOCHEMICAL ANALYSIS OF LIVER FUNCTIONS

The average biochemical activities in mice belonging to each group (Table 3.1) prove normal levels of both ALT and direct bilirubin indicating normal liver functions.

TABLE 3.1 Representation of Serum ALT Activity and Concentration of Direct Bilirubin in Ag or Ag/Co Mediated PTT

Group	ALT (U/I)	Serum Direct Bilirubin (mg/dL)
Normal Control	9.30653	0.1543
Tumor Control	3.51667	0.1584
Light Control	6.24131	0.2592
Dark Control (Ag)	2.66522	0.1872
Dark Control (Ag/Co)	3.0058	0.1872
Treated (Ag)	2.83551	0.1239
Treated (Ag/Co)	2.15435	0.1152
Normal Range	Up to 14 U/I	Up to 0.25

3.3.4 HISTOLOGICAL ANALYSIS

Histological sections from normal control mice showed striated skeletal muscles, with no evidence for the presence of any cancer cells (Figure 3.5A). In the light control mice group, at which tumors have been exposed to the monochromatic blue LED without pre-injection with either nanoparticles, inflammatory cell infiltration was recognized in the striated muscles (Figure 3.5B). In the tumor control mice, the cancer tissue is recognized by the presence of giant cells with mitotic and hyperchromatic nuclei replacing the striated muscles and the subcutaneous tissue. These findings indicate the successful induction of solid tumor formation (Figure 3.5C–D).

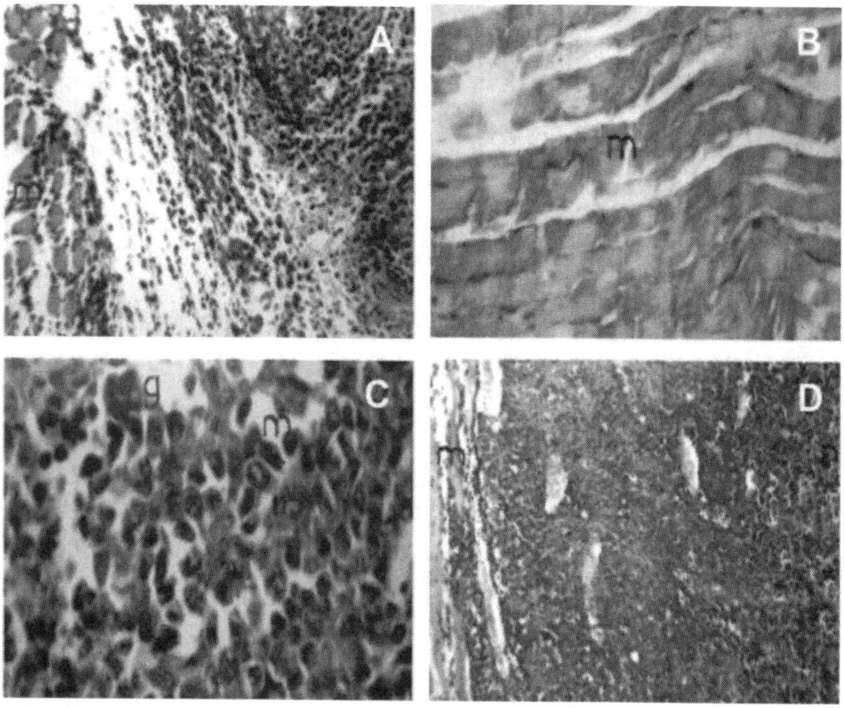

FIGURE 3.5 Normal histological features of striated muscle in a normal control mouse (A). Histological section in the light control mouse showing inflammatory cells among the muscle fibers "m" with intact neoplastic cells "n" in the adjacent area (B). Skeletal muscles and subcutaneous tissue of tumor control mouse showing groups of cancer cells "n" replacing striated muscles "m" (C), with giant tumor cells "g" and mitotic "m" hyperchromatic nuclei "h" (D).

On the other hand, the dark control tumor tissues, which have been injected with either AgNPs (Figure 3.6A), or Co/AgNPs (Figure 3.6C) without light exposure, showed the presence of necrobiotic tumor cells, as well as intact cancer cells in the subcutaneous tissue of those injected with AgNPs, but more intact tumor cells with less necrobiosis in tumors injected with Co/AgNPs. Treated tumors with AgNPs or Co/AgNPs mediated PTT via 1 h exposure to monochromatic blue light post NPs injection exhibited high degree of necrosis as well as some intact tumor cells in the subcutaneous tissue of the AgNPs treated tumors (Figure 3.6B), and large necrotic areas with some necrobiosis in the subcutaneous tissue and few intact cancer cells (Figure 3.6D).

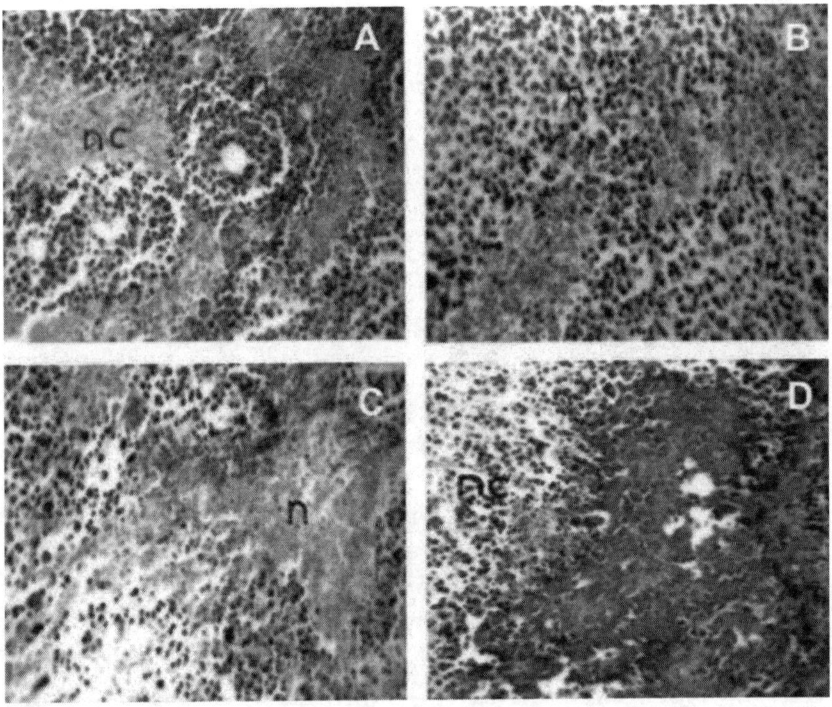

FIGURE 3.6 Histological sections from tumors injected with Ag (A & B) or Co/Ag (C & D). The dark control tumors injected with Ag show necrobiotic "nc" as well as intact cancer cells in the subcutaneous tissue (A). Treated tumors with Ag+1 h light exposure, contain high degree of necrosis "nc" as well as some intact tumor cells "n" in the subcutaneous tissue (B). Dark control tumors injected with Co/Ag have large number of intact tumor cells "n" and lower degree of necrobiosis "nc" in the subcutaneous tissue (C). Treated tumors with Co/Ag+1 h light exposure, have large necrotic areas "nc" with some necrobiosis "nb" in the subcutaneous tissue and few intact cancer cells (D).

3.4 DISCUSSION

As little is known about their biological applications in cancer treatment, and relying on the fact of being novel candidates providing high thermal effect, this study was directed towards *in vivo* investigation of the photothermal tumor ablation using silver and cobalt core silver shell nanoparticles in an albino mouse Ehrlich solid tumor model.

In order to detect the photothermal efficacy of AgNPs and Co/AgNPs in solid tumor ablation, several parameters have been considered, including reduction in tumor volume and histopathological modifications post-treatment. The advantages of such treatment modality have been evaluated by investigating the side effects that might appear during or after application of the therapeutic program, such as; the general body temperature, and/or the animal physiological conditions.

Normal tissues have tight, continuous vessel walls with pores that are approximately 9 – 50 nm diameter. Large pores (larger than 50 nm diameter) are 100 – 1000 times fewer in number than the 9 nm pores, making it easier for small molecules to penetrate all types of tissues than large molecules including polymers that do so very slowly. Tumor tissues, which have discontinuous capillary walls is used to suspend polymeric materials from the body allowing particles less than 100 nm diameter, to penetrate easily (Gümüşhan et al., 2013). Therefore, once nanoparticles penetrate into tumor tissue, they cannot be easily removed. In addition, tumors exhibit enhanced penetration and retention effect for particles of 50 − 100 nm (Hou et al., 2009). Thus, hybrid nanoparticles from 50 to 100 nm are well suited to target tumors for imaging and therapeutic purposes.

On the other hand, tumor tissues have discontinuous capillary walls that have no basal lamina and have a large number of pores of almost 100 nm diameter allowing particles smaller than this size to penetrate easily. Such features lead to simple diffusion, rapid penetration and high accumulation rate of 50–100 nm particles ().

Alternatively, tumor tissues are known to lack lymphatic system for eliminating lipophilic and polymeric materials from them, meaning that particles cannot be rapidly eliminated after they enter the tumor cells (Gümüşhan et al., 2008). These features render the tumor-enhanced penetration, as well as retention to be effective for 50–100 nm particles. Therefore, long-circulating nanoparticles with the appropriate properties and size have a big chance to encounter leaky tumor capillaries and navigate into the tumor tissue.

After light irradiation of AgNPs and Co/AgNPs treated mice, a black color could be detected allover the tumor area. This may be due to the oxidation of the silver metal to form silver oxide with its specific black color (Taylor and Granger, 1983). The shrinkage in tumor size was mainly due to the hyperthermia that resulted from activation of the injected nanoparticles by the monochromatic LED of wavelength 460 nm. This hyperthermia can induce necrotic cell death via heat shock protein (HSP) expression, which also induces antitumor immunity (Dvořák, 1998). The other scenario of cell death by nanoparticles is associated with oxidative stress, endogenous reactive oxygen species (ROS) production and depletion of the intracellular oxidative pool (Yin et al., 2002; Ito et al., 2006).

In the dark control group of mice that has been injected with either AgNPs or Co/AgNPs, some necrotic tissue appeared at day 6 in the center of the tumor mass, however, the tumors continued to grow. The central necrosis could merely be due to the combined effect of poor nutrition supply at the center of the rapidly growing tumor mass, and the presence of foreign objects, which are the nanoparticles. The continuous growth of the tumors suggests that the presence of nanoparticles without being activated with the specific wavelength has no direct effect on the inhibition of tumor growth.

On the other hand, animals in the light control group did not show any suppression in tumor growth by exposure to LED without being injected with either type of the nanoparticles. This indicates and supports the idea that reduction in tumor volume was only due to the combined effect of the monochromatic light in producing the thermal cytotoxic effect upon activation of the used nanoparticles.

Our findings come in agreement with other studies that proposed cancer treatment using magnetite cationic liposomes and hydroxyapatite nanoparticles, successively resulting into local hyperthermia when exposed to inductive heater with high frequency alternating magnetic field (Foldbjerg, 2009).

A comparison between variations in body temperature between the treated mice and the anesthetized control mice indicates only 2.5°C decrease, which is still in the normal range of the control mice. Those treated mice did not show any signs of dehydration during light exposure and all mice survived for 30 days post-treatment. As a result, the exposure is considered safe when operated under general anesthesia, since the final stable elevated core body temperature is within the range of normal body temperature. These findings are in accordance with the study done by Hou et al. (2009).

The normal levels of biochemical parameters prove that the metabolism of both AgNPs and Co/AgNPs mediated PTT does not affect the liver functions or cause hepatotoxicity. This might be a result of the local injection of the nanoparticles in the tumor mass, decreasing the amount of NPs in the systemic circulation. The short period between nanoparticles injection and light exposure was not enough for nanoparticles elimination by tumor cells to the bloodstream.

Histological examination of the treated mice indicated that the AgNPs mediated photothermally treated tumors showed high number of necrotic cells with some intact tumor cells in the subcutaneous tissue, while the Co/AgNPs mediated photothermally treated tumors have a wider range of necrosis with some areas of necrobiosis in the subcutaneous tissue and a very limited number of intact cancer cells. Both groups of treated mice proved the presence of more necrotic tissue as compared to the dark control tumors that have been injected with either AgNPs or Co/AgNPs. It is worth mentioning that the level of necrosis in the treated tumors was higher in case of Co/AgNPs than AgNPs, while in the dark control samples, the Co/AgNPs injected tumors have more intact cells than in AgNPs injected tumors.

Our previous *in vitro* study on AgNPs and Co/AgNPs mediated photothermal therapy on Hep-2 laryngeal carcinoma cell line toxicity, and their genetic outcomes on blood lymphocytes could in part reflect the mechanism of tumor cells killing at the in vivo level (Gomaa et al., 2015).

3.5 CONCLUSIONS

Results of this work concluded that Co/AgNPs have less dark toxicity, but higher light toxicity than the AgNPs. Accordingly, Co/AgNPs is less toxic to tissue that is not exposed to light. This proposes specific tumor cells killing by maximizing the duration of tumor cells exposure to the LED and minimizing it to adjacent tissues in which nanoparticles might diffuse from the injected tumor mass.

ACKNOWLEDGEMENT

We thank Marwa Ramadan for the preparation of the Ag & Ag/Co nanoparticles at the laboratories of the National Institute for Laser Enhanced Sciences (NILES), Cairo University. Our thanks are also directed to Mohamed Ghaly at the Faculty of Pharmacy and Biotechnology for his technical assistance. The authors gratefully acknowledge the financial support of the "Center for Specialized Studies and Programs (CSSP) – Bibliotheca Alexandrina).

Conflict of Interest

The authors report no conflicts of interest in this work. This work has an international PCT publication No. WO 2011/110186 A1.

KEYWORDS

- **Ag and Co/Ag nanoparticles**
- **photo-thermal therapy**
- **solid tumor ablation**

REFERENCES

Asane, G. S. "Mucoadhesive Gastrointestinal Drug Delivery System." Pharmainfo.net 2007, 5(6).

Chen, Minghai, Lian Gao. "Synthesis and characterization of Ag nanoshells by a facile sacrificial template route through in situ replacement reaction." *Inorganic Chemistry* 2006, 45(13), 5145–5149.

Dvořák, Bohuslav nská, Debra, L. McWilliam, Catherine, S. Williams, Travis Higdon, Marie Zákostelecká, and Otakar Koldovský. "The expression of epidermal growth factor and transforming growth factor-α mRNA in the small intestine of suckling rats: organ culture study." *FEBS Letters* 1998, 435(1),119–124.

Foldbjerg, Rasmus, Ping Olesen, Mads Hougaard, Duy Anh Dang, Hans Jürgen Hoffmann, Herman Autrup. "PVP-coated silver nanoparticles and silver ions induce reactive oxygen species, apoptosis and necrosis in THP-1 monocytes." *Toxicology Letters* 2009, 190(2), 156–162.

Gabriel, J. The Biology of Cancer; Second Edition 2007, Central South Coast Cancer Network.

Gümüşhan, Hatice, Davut Musa. "Effect of adriamycin administered via different routes on Ehrlich ascites tumor cells." *IUFS J Biol* 2008, 67, 49–54.

Gümüşhan, Hatice, Davut Musa. "Effect of adriamycin administered via different routes on Ehrlich ascites tumor cells." *IUFS J Biol* 2008, 67, 49–54.

Haberl, Nadine, Stephanie Hirn, Alexander Wenk, Jörg Diendorf, Matthias Epple, Blair, D. Johnston, Fritz Krombach, Wolfgang, G. Kreyling, Carsten Schleh. "Cytotoxic and proinflammatory effects of PVP-coated silver nanoparticles after intratracheal instillation in rats." *Beilstein Journal of Nanotechnology* 2013, 4(1), 933–940.

Hou, Chun-Han, Sheng-Mou Hou, Yu-Sheng Hsueh, Jinn Lin, Hsi-Chin Wu, and Feng-Huei Lin. "The in vivo performance of biomagnetic hydroxyapatite nanoparticles in cancer hyperthermia therapy." *Biomaterials* 2009, 30(23), 3956–3960.

Iman Gomaa, Samarth Bhatt, Thomas Liehr, Mona Bakr, Tarek El-Tayeb. Ag and Co/Ag nanoparticles cytotoxicity and genotoxicity study on Hep-2 and blood lymphocytes. Chapter 2, Chemical Technology- Key development in applied chemistry, Biochemistry and Material Science. Apple Academic Press, 2015, 13–29.

Ito, Akira, Hiroyuki Honda, and Takeshi Kobayashi. "Cancer immunotherapy based on intracellular hyperthermia using magnetite nanoparticles: a novel concept of "heat-controlled necrosis" with heat shock protein expression." *Cancer Immunology, and Immunotherapy*, 2006, 55(3), 320–328.

Kikumori T, Kobayashi T, Sawaki M, Imai, T. Anti-cancer effect of hyperthermia on breast cancer by magnetite nanoparticle-loaded anti-HER2 immunoliposomes. *Breast Cancer Res Treat.* 2009, 113(3), 435–441.

Liau, S. Y., et al. "Interaction of silver nitrate with readily identifiable groups: relationship to the antibacterialaction of silver ions." *Letters in Applied Microbiology* 1997, 25(4), 279–283.

McNeil, Scott, E. "Nanotechnology for the biologist. "Journal of Leukocyte Biology" 2005, 78(3), 585–594.

Nam, Y. S., Kang, H. S., Park, J. Y., Park, T. G., Han, S. H., Chang, I. S. New micelle-like polymer aggregates made from PEI-PLGA diblock copolymer s: micellar characteristics and cellula r uptake. Biomaterials 2003, 24, 2053–2059.

Osaka, Tetsuya, et al. "New trends in nanoparticles: syntheses and their applications to fuel cells, health care, and magnetic storage." *Israel Journal of Chemistry* 2008, 483–4, 333–347.

Praetorius, Natalie, P., Tarun, K. Mandal. "Engineered nanoparticles in cancer therapy." *Recent Patents on Drug Delivery & Formulation* 2007, 1(1), 37–51.

Ruddon, R. W. "Characteristics of human cancer. "Cancer Biology". Fourth Edition, Oxford University Press, New York, 2007, 3–16.

Schneidewind, Henrik, Thomas Schüler, Katharina, K. Strelau, Karina Weber, Dana Cialla, Marco Diegel, Roland Mattheis, Andreas Berger, Robert Möller, and Jürgen Popp. "The morphology of silver nanoparticles prepared by enzyme-induced reduction." *Beilstein Journal of Nanotechnology* 2012, 3(1), 404–414.

Taylor, A. E., Granger, D. N. *Fed Proc* 1983, 42(8), 2440–5.

van der Zee, J. Heating the patient: a promising approach? *Ann Oncol* 2002, 13, 1173–1184.

Yin, Yadong, Zhi-Yuan Li, Ziyi Zhong, Byron Gates, Younan Xia, Sagar Venkateswaran. "Synthesis and characterization of stable aqueous dispersions of silver nanoparticles through the Tollens process." *J. Mater. Chem.* 2002, 12(3), 522–527.

CHAPTER 4

INTEGRATING BIOINFORMATICS IN ALTERNATIVE MEDICINE PRACTICE: AN INSIGHT

I. R. OVIYA

CONTENTS

ABSTRACT

The knowledge from past and present holds our future. Alternative medicine is age-old tradition and proven skill which is presently entering the mainstream holding promises for combating various dreadful diseases in which our modern medicines have failed. Though the traditional medicine has been developed by more of observation, practice and skill rather than scientific proven mechanism, still, the medicinal effectiveness is too broad and effective to be ignored. Recent advances in holistic medicinal research are towards uncovering the molecular mechanism of inhibition. Advances in biology, biotechnology has greatly improved our understanding of the human system and mode of drug-target interaction. To aid with these technologies, computational approach termed bioinformatics plays an important role. Genomics, proteomics, transcriptomics, data mining, text mining, network construction, expression studies, etc. are some areas of bioinformatics which when interconnected interprets various important information on the medical aspect and opens up challenges and solutions towards personalized medicine and drug discovery.

4.1 INTRODUCTION

The age-old practice of traditional medicine is gaining momentum today. The traditional and alternative medicinal practices like ayurveda, siddha, unani, kampo, traditional Chinese medicine, etc., have promising therapeutic values. They constitute the group of evidence-based medicine. Though not scientifically proven then, but the practical knowledge of our forefathers have unbelievably exceeded today's scientific practice when the whole foundation was based upon personalized medicinal concept or 'Prakriti' which is also mentioned in our ancient text as follows (Chatterjee and Pancholi, 2011; Patwardhan and Khambholja, 2011):

> *'Every individual is different from another*
> *and hence should be considered as a different entity.*
> *As many variations are there in the Universe, all are seen in Human being'.*

With the advancement in biological science, the root cause of all human diseases is mapped to the genomic and phenotypic variations, which are important to understand for personalized medicine concept. Prakriti-based medicine is based on three doshas, namely, vata, pitta and kapha explaining the different conditions of the body.

4.2 WHAT IS ALTERNATIVE MEDICINE?

Alternative medicine, which includes ayurveda, siddha, unani, kampo, traditional Chinese medicine, to name a few, is a substitute for present conventional medicine, namely Allopathy. They are practiced with traditional knowledge for healing in different parts of the world. The ancestral medicinal knowledge documented by various scholars of their era is now being used to gain knowledge for alternative medicinal practice.

4.2.1 PROS AND CONS

1. Alternative medicines are based on evidence-based practice since ancient times. Thus, it has a strong foundation for sustaining its practice.
2. They are believed to posses lesser side effects.
3. Alternative medicine is based on personalized medicine concept. Hence, drug selection and dosage quantity is based on the prakriti or tridosha of a person under treatment.
4. Alternative medicines are believed to cure some of the most menacing disease of present era which the present conventional medicine is unable to cure.
5. Presently there is no standardization of herbal medicine which compromises on the purity of the medicine.
6. Still various practices of the tribal/communities are undisclosed. Hence, very little is known of different traditional remedies practiced today.
7. Scientific research is still lagging in understanding the mechanism of action of the herbal formulations and its synergistic activity in the biological system.

4.3 WHAT IS BIOINFORMATICS?

Bioinformatics is an interdisciplinary area for analyzing the biological systems with a computational approach. The area of bioinformatics has diversified into many sister areas of research namely, genomics, proteomics, metabolomics, systems biology, pharmacogenomics, etc. More recently few new areas seems to emerge like ayugenomics (Patwardhan and Khambholja, 2011), reverse pharmacogenomics (Anonymous, 2004) and pathogenomics (Kumar, 2007).

4.3.1 GENOMICS

Genomics is the study of sequence, structure and function of genome. The completion of the human genome project held high hopes of the scientists in un-deciphering the genetic puzzle in humans in medical perspective. Breakthrough in establishing basis of individual differences with VNTRs, SNPs, EST and other genomic elements gave hope towards a personalized medication approach. However, the science of human body is too complex to understand and apply these concepts to the whole.

4.3.2 PROTEOMICS

Proteomics is the study of structure and function of proteins. In recent times, there has been a boom in structure elucidation of various proteins and its mode of action. These structures give us better understanding of the protein behavior with different molecules. The identification of the active site and the active residues are the main components into probing the possible drug-target interaction.

4.3.3 TRANSCRIPTOMICS

Transcriptomics is the study of expression level of different proteins in a given cell population. It is also referred as expression profiling and is based on microarray technology. This technology is used to study the differences in the expression of different genetic elements in more than one condition such as normal *vs*. disease conditions.

4.3.4 METABOLOMICS

The study of various intermediate metabolites in cellular processes and decoding the chemical fingerprints is known as metabolomics.

4.3.5 SYSTEMS BIOLOGY

Systems biology focuses on integrating the data from genomics, proteomics, transcriptomics and metabolomics. It focuses on computational model building using the data involving metabolic networks or cell signaling networks. Network building is an integral part of systems biology.

4.3.6 PHARMACOGENOMICS

Pharmacogenomics aims towards personalized medicine approach. It involves technology combining genomics, proteomics, metabolomics, etc. to analyze the individual's genetic makeup affecting his drug response.

This integrative field is still in its infancy with respect to understanding of complexity of the biological entities. However, the advances made in the sister areas can't be ignored.

4.4 BIOINFORMATICS AND ALTERNATIVE MEDICINE: THE PRESENT SCENARIO

The promises of the computational models for studying the biological systems is not new but the advent of the technology post the completion of the human genome project has been huge. High throughput sequencing data, microarray and expression data compiled systematically in a database is a boon for the modern biologist for extracting the experimental knowledge and its application henceforth. The present day approaches in alternative medicine is mainly focuses on QSAR studies, molecular modeling, docking, dynamic and other validation studies. These studies are mainly focused on the application of the developed strategies. The prospects of bioinformatics is not only limited to the usage of available data and softwares but to develop more robust network models to comply with the dynamicity of the biological systems. Human system is complex as well as variable. The earliest scientific approach was restricted to understand the human complexity. Recent studies have to be extended to deal with the variability in genome induced due to physiological and biological changes. The computational biology plays an integral role in bioinformatics in developing the softwares for analysis. The efficiency of the computational techniques is dependent on the mathematical models implemented using programming languages, which can deal with both complexity and variability as mentioned.

4.4.1 DATABASES, TEXT MINING METHODS/LITERATURE BASED STUDIES

The information on the traditional medicine is scattered and the present day need is to gather them and present it to the researchers in a more organized manner. Various kind of information are available in different databases which minimize our searing work and allow us to effective utilize the information for further analysis. The traditional practices and usages that have been documented in different books, scriptures and sometimes they have not been documented but have been in practice with the group or tribe for a long time. Such kind of information is highly valuable and cannot be harnessed at one place. Hence the research work comprising of field visit, collection of traditional information and documentation in modern form (databases) is important to give other researcher a peep into the data and to carry out the further pharmacognostical or phytochemical work. Medicinal pharmacogenomics base is in the information available likewise. Similarly, along with documentation, web search needs to be more robust and specific. Text mining approaches improves not only efficiency of search but also saves time. Few databases for the reference of the reader are mentioned in Table 4.1.

TABLE 4.1 List of Databases for References

Databases	Information	Reference
Supernatural	Available natural compounds	Dunkel et al. (2006)
TCM-ID	Herbal ingredient, structure, function, therapeutic details	Xue et al. (2013)
TCMGeneDIT	Association information about TCM	Fang et al. (2008)
HIT	Herbs and their protein target information	Hao et al. (2011)
SWEETLEAD	Approved drugs, regulated chemicals, herbal isolates for CADD	Novick et al. (2013)
INpacdb	Indian plant anticancer compounds	Vetrivel et al. (2009)
NAPROC-13	Indian medicinal plants	Lopez-Perez et al. (2007)

4.4.2 EXPRESSION STUDIES

Microarray analysis, spectroscopic methods and other high throughput technology have lead to the increase in the pharmacological understanding of the disease conditions. The differential gene, protein expression can help us to identify the candidate gene and thus help to identify the drug target. The expression data can be accessed by various public databases like GEO, ArrayExpress, etc.

4.4.3 TRADITIONAL CHINESE MEDICINE (TCM)

The worldwide acceptance of TCM makes it a promising source of experimental studies. The concept of personalized medicine is not new and that it is being practiced for thousands of years. The basis of TCM practice lies in Zheng i.e. personalized identification and classification of the symptoms and treating it individually. Zheng describes the overall physical status of the human body based on the genotypic and phenotypic differences arising due to heredity, SNPs, environmental variations and nutrition wherein the patients can classified into characteristic hot or cold Zheng. This has given our modern day researchers to probe into the effectiveness of TCM and its underlying mechanism. Various research papers suggest *in-silico* approach to solve the puzzle of TCM. The sequencing data, text mining resources, pharmacogenomics, microarray expression analysis, etc. all gave a reason to systematically investigate TCM further (Song et al., 3013). More recently systems biology and network pharmacology is the principle choice of carrying out study designs on TCM. Network pharmacology integrates the available chemical and metabolomic knowledge derived from experimental results for drug-target interaction or network studies in a computer-simulated model mimicking the living systems. The mode of actions and the scientific basis behind TCM's principle is still unclear. There is a curiosity to unveil the principles behind TCM life systems. Kang et al. developed novel entropy based models like acquired life entropy, acquired life entropy flow and acquired life entropy production using the experimental data and has drawn comparative lines with the current TCM principles like zheng, *qi*, yin-yang, etc. (2008). The computational approach is also used to study molecular mechanisms like miRNA and siRNA interaction, glycolysis targeting, histone modifications, and DNA methylation (Hsieh et al., 2011; Hsieh et al., 2013, Wang et al.,

2012). These mechanisms are particularly important to understand as they serve as the basis for the drug-target interaction studies. The present day drugs are based on 'one drug fits all' theory, which may not be a pharmacologically correct term for treating a disease. Network pharmacology is being used (Li et al., 2012) for multi-compound drug discovery. Diseases like cardiovascular disease, HIV, etc. that are being treated using TCM are being probed by computational methods into its mode of action (Hu et al., 2010). Various mathematical models have been suggested for network construction between herbs/natural products and their targets (Li et al., 2012; Zhao et al., 2013; Li, 2009; Li et al., 2010, 2011; Fang et al., 2013). The concept of network pharmacology is explained in detail in various review articles (Li et al., 2011; Fang et al., 2013).

The various steps involved in bioinformatics approaches are:

1. Text mining for collection of literature;
2. Data mining of the various available resources from databases. For example, 2D and 3D structure of compounds from chemspider, pubchem, etc.;
3. Understanding the problem, finding loopholes, and study design;
4. QSAR studies;
5. Target identification and target modeling;
6. Network building, interpretation, visualization, and pathway-enrichment analysis;
7. Genome-wide association studies;
8. Evolutionary analysis;
9. Docking and molecular simulation.

4.5 CONCLUSION

No science can justify independently its role in a biological system. Different areas of science have to come together for sharing scientific concepts and knowledge. Bioinformatics plays an important role in understanding biological mechanism and unveiling the underlying genomic entities responsible for causing diseases. Alternative and complementary medicine is presently gaining momentum. Ethnomedicine, phytochemistry, pharmacology of the plants and its mechanism of action are being studied by using various molecular techniques and bioinformatics tools.

The need of the hour is to fasten the research by using *in silico* methods coupled with other biological techniques to scientifically validate the usage of the herbal medicines. Also there is a need for the researchers to gain the trust of the tribal people so as to utilize their knowledge. Study designs should be more comprehensive to fill the void in the medical knowledge.

KEYWORDS

- **alternative medicine**
- **bioinformatics**
- **databases**
- **genomics**
- **metabolomics**
- **proteomics**

REFERENCES

Anonymous. Ayurveda and drug discovery. Current Science 2004, 86(6), 754.

Patwardhan, B., Khambholja, K. Drug Discovery and Ayurveda: Win-Win Relationship Between Contemporary and Ancient Sciences, Drug Discovery and Development - Present and Future, Kapetanović, I., Ed.; InTech, 2011, p 528.

Chatterjee, B., Pancholi, J. *Prakriti*-based medicine: A step towards personalized medicine. Ayu. 2011, 32(2), 141–146.

Dunkel, M., Fullbeck, M., Neumann, S., Preissner, R. SuperNatural: a searchable database of available natural compounds. Nucleic Acids Res. 2006, 34, D678–D683.

Fang, Y. C., Hsuan-Cheng Huang, H. C., Chen, H. H., Juan, H. F. TCMGeneDIT: a database for associated traditional Chinese medicine, gene and disease information using text mining. *BMC Complementary and Alternative Medicine* 2008, 8, 58.

Fang, Z., Zhang, M., Yi, Z., Wen, C., Qian, M., Shi, T. Replacements of Rare Herbs and Simplifications of Traditional Chinese Medicine Formulae Based on Attribute Similarities and Pathway Enrichment Analysis. Evidence-Based Complementary and Alternative Medicine 2013, 9 pages.

Hsieh H. Y., Chiu, P. H., Wang. S. C. Epigenetics in Traditional Chinese Pharmacy: A Bioinformatic Study at Pharmacopoeia Scale. Evidence-Based Complementary and Alternative Medicine 2011, 10 pages.

Hsieh H. Y., Chiu, P. H., Wang. S. C. Histone modifications and traditional Chinese Medicinal. BMC Complementary and Alternative Medicine 2013, 13, 115.

Hu, J. Z., Bai, L., Chen, D. G., Xu, Q. T., William, M. Southerland. Computational Investigation of the Anti-HIV Activity of Chinese Medicinal Formula Three-Huang Powder. Interdiscip Sci. 2010 June; 2(2), 151–156.

Kang, G. L., Shao Li, S., Zhang, J. F. Entropy-Based Model for Interpreting Life Systems in Traditional Chinese Medicine. eCAM 2008, 5(3), 273–279.

Kumar, D. From evidence-based medicine to genomic medicine. Genomic Med. 2007, 1, 95–104.

Li, B., Xu, X., Wang, X., Yu, H., Li, X., Tao, W., Wang, Y., Yang, L. A Systems Biology Approach to Understanding the Mechanisms of Action of Chinese Herbs for Treatment of Cardiovascular Disease. Int. J. Mol. Sci. 2012, 13, 13501–13520.

Li, J., Lu, C., Jiang, M., Niu, X., Guo, H., Li, L., Bian, Z., Lin, N., Lu, A. Traditional Chinese Medicine-Based Network Pharmacology Could Lead to New Multicompound Drug Discovery. Evidence-Based Complementary and Alternative Medicine 2012, 11 pages.

Li, S. Network Systems Underlying Traditional Chinese Medicine Syndrome and Herb Formula. *Current Bioinformatics*, 2009, 4, 188–196.

Li, S., Zhang, B., Jiang, D., Wei, Y., Zhang, N. Herb network construction and co-module analysis for uncovering the combination rule of traditional Chinese herbal formulae. BMC Bioinformatics 2010, 11(Suppl 11), S6.

Li, S., Zhang, B., Zhang, N. Network target for screening synergistic drug combinations with application to traditional Chinese medicine. BMC Systems Biology 2011, 5(1), S10.

López-Pérez, J. L., Therón, R., Olmo, D. E., Díaz, D. NAPROC-13: a database for the dereplication of natural product mixtures in bioassay-guided protocols. Bioinformatics 2007, 23(23), 3256–3257.

Novick, P. A., Ortiz, O. F., Poelman, J., Abdulhay, A. Y., Pande, V. S. SWEETLEAD: An in silico database of approved drugs, regulated chemicals, and herbal isolates for computer-aided drug discovery. PLoS ONE 2013, 8(11), e79568.

Song, Y. N., Zhang, G. B., Zhang, Y. Y., Su, S. B. Clinical Applications of Omics Technologies on ZHENG Differentiation Research in Traditional Chinese Medicine Evidence-Based Complementary and Alternative Medicine 2013, 11 pages.

Vetrivel, U., Subramanian, N., Pilla, K. InPACdb—Indian plant anticancer compounds database. Bioinformation. 2009, 4(2), 71–74.

Wang, Z., Wang, N., Chen, J., Shen, J. Emerging Glycolysis Targeting and Drug Discovery from Chinese Medicine in Cancer Therapy. Evidence-Based Complementary and Alternative Medicine 2012, 13 pages.

Xue, R., Fang, Z.,, Zhang, M., Yi, Z., Wen, C., Shi, T. TCMID: Traditional Chinese Medicine integrative database for herb molecular mechanism analysis. Nucleic Acids Res. 2013, 41, D1089–95.

Ye, H., Ye, L., Kang, H., Zhang, D., Tao, L., Tang, K., Liu, X., Zhu, R., Liu, Q., Chen, Y. Z., Li, Y., Zhiwei Cao, Z. HIT: linking herbal active ingredients to targets. Nucleic Acids Res. 2011, 39, D1055–D1059.

Zhao, M., Zhou, Q., Ma, W., Wei, D. Q. Exploring the Ligand-Protein Networks in Traditional Chinese Medicine: Current Databases, Methods, and Applications. Evidence-Based Complementary and Alternative Medicine 2013, 15 pages.

CHAPTER 5

RECENT ADVANCES IN HIV/AIDS

MOHAMMAD HUSAIN and POONAM GUPTA

CONTENTS

5.1 INTRODUCTION

Acquired Immune deficiency Syndrome (AIDS), the dreaded disease was discovered in 1981. It came into highlight when a large number of young homosexual men succumbed to unusual opportunistic infections and rare malignancies (CDC, 1981; Greene, 2007). The principal causative agent is human immunodeficiency virus (HIV-1) that belongs to retrovirus group of viruses (Barre-Sinoussi et al., 1983; Gallo et al., 1984; Popovic et al., 1984). Another causative agent, HIV-2 is rarely found and is less pathogenic. Both the viruses are the result of multiple cross-species transmissions of simian immunodeficiency viruses (SIVs) that naturally infect African primates. These cross-species transmission events, involving transmission of SIVcpz (SIV strain infecting chimpanzee) from south-eastern Cameroon to humans led to emergence of HIV-1 group M (Paul et al., 2011).

HIV-1 spreads by sexual, percutaneous and perinatal routes. It is known that 80% of adults acquire infection through sexual route and thus categorized as sexually transmitted disease (Hladik and McElrath, 2008; Cohen et al., 2011).

5.2 HIV-1

HIV-1 has been found to be originated from Simian Immunodeficiency Viruses (SIVs) infecting chimpanzees and gorilla. The molecular epidemiologic data has clearly indicated that HIV-1 evolved with the *Pan troglodytes troglodytes* subspecies of chimpanzee and was present in that subspecies for centuries (Gao et al., 1999). The most likely mechanism of transmission of HIV-1 from chimpanzees to humans was by contamination of a person's open wound with the infected blood of a chimpanzee, probably when the chimpanzee was being butchered for the purpose of consumption (Weiss and Wrangham, 1999). Chimpanzees have traditionally served as a source of nutrition for humans in certain parts of sub-Saharan Africa. Any of a number of mutations in the viral genome that would have allowed successful transmission of the virus from chimpanzee to humans probably took place intermittently over the centuries (Gao et al., 1999).

A major characteristic of human immunodeficiency viruses (HIV) is their high genetic variability. This led to the classification of HIV-1 into three groups based on nucleotide sequence analyses: M (major), O (outlier), and N (non-M/non-O) and recently found P. Each of the lineages resulted from an independent cross-species transmission event (Paul et al., 2011). Group M was the first to be discovered and represents the pandemic form of HIV-1; it has been reported to infect millions of people worldwide and is prevalent globally. Group O was discovered in 1990 and is much less prevalent than group M (De et al., 1990; Gurtler et al., 1994). It represents less than 1% of global HIV-1 infections and is largely restricted to Cameroon, Gabon, and neighboring countries (Mauclere et al., 1997; Peeters et al., 1997) and also less prevalent than group M. Group N was identified in 1998 (Simon et al., 1998), and is even less prevalent than group O, as so far been detected only in Cameroon (Vallari et al., 2010). Finally, group P was discovered in 2009 in a Cameroonian woman living in France (Plantier et al., 2009). Recently, another woman from Cameroon only was reported with group P virus (Vallari et al., 2011). The most recent nomenclature of HIV-1 divides group M into subtypes A-D, F-H, and J-K, with further subdivision into subtypes F1 and F2. In addition, there are several groups of circulating recombinant forms (CRFs) of HIV-1 (Robertson et al., 2000). Broadly subtype B is exclusively dominant in North and South America, Western Europe and Australia, subtype C in India, Southern Africa and Ethiopia, subtype A in Eastern Europe, Central Asia and some parts of African continent. In terms of global predominance subtype C accounts for almost 55% of infection while subtypes A, B, D and G have been reported to account for 12, 10, 3 and 6% of infections respectively (Mannar et al., 2006). The Predominance of various subtypes in different geographical location is given in the Table 5.1.

TABLE 5.1 Geographic Distribution of HIV-1 Groups and Various Subtypes

Groups Distribution	Sub-types	Geographic/Country
M	A	East Africa, West Africa, Central Africa, Eastern Europe and Central Asia
	B	North America, South America, East Africa, Central Africa, North Africa/Middle, East, Europe, Australia, New Zealand, Japan, China, Korea, Philippines and Malay Peninsula
	C	India, Brazil, South Africa, East Africa, Nepal and China
	D	India, Brazil, South Africa, East Africa, Eastern Europe and Central Asia

TABLE 5.1 *(Continued)*

Groups Distribution	Sub-types	Geographic/Country
	F	Central Africa, West Africa, Latin America, Caribbean and North America
	G	West Africa, Central Africa, North Africa, Middle East, Eastern Europe, Taiwan and Korea
	H	Central Africa, Eastern Europe and Central Asia
	J	Central Africa, West Africa
	K	Cameroon, DR Congo
N	Nil	West Africa (Cameroon, Niger, Gabon, Nigeria, Senegal, and Tongo), France and Norway
O	Nil	Cameroon
P	Nil	Cameroon

5.3 HIV-2

Another morphologically similar and antigenically distinct virus was discovered in 1986 (Clavel et al., 1986) and termed HIV-2. It is largely restricted in West Africa, with highest prevalence rates recorded in Guinea-Bissau and Senegal (de et al., 2008). Most of the patients infected with HIV-2 do not progress to AIDS (Rowland-Jones and Whittle, 2007). The virus is known to originate from a sooty mangabey strain (Hirsch et al., 1989). HIV-2 has eight different lineages termed as groups A-H. Group A and B are known to infect human population to an appreciable degree. Group A is prevalent in western Africa whereas group B predominates in Cote d'Ivoire (Pieniazek et al., 1999; Ishikawa et al., 2001). All other groups were initially found in single individuals. Of these, groups C, G and H are known to originate from SIVsmm strains from Cote d'Ivoire, group D from SIVsmm strains from Liberia, and groups E and F from Sierra Leone (Gao et al., 1992; Chen et al., 1996, 1997; Santiago et al., 2005).

5.4 STRUCTURE OF HIV-1

Human immunodeficiency virus (HIV) is a lentivirus (a member of the retrovirus family) that leads to acquired immunodeficiency syndrome (AIDS), a condition in humans in which the immune system begins to fail, leading to life-threatening opportunistic infections (*Coffin et al., 1986*). HIV is different in structure from other retroviruses. It is roughly spherical (*McGovern et al., 2002*) with a diameter of about 120 nm, around 60 times smaller than a red blood cell, yet large for a virus. It is composed of two copies of positive single-stranded RNA that codes for the virus' nine genes enclosed by a conical capsid composed of 2,000 copies of the viral protein p24 (Figure 1.1 and Table 5.2).

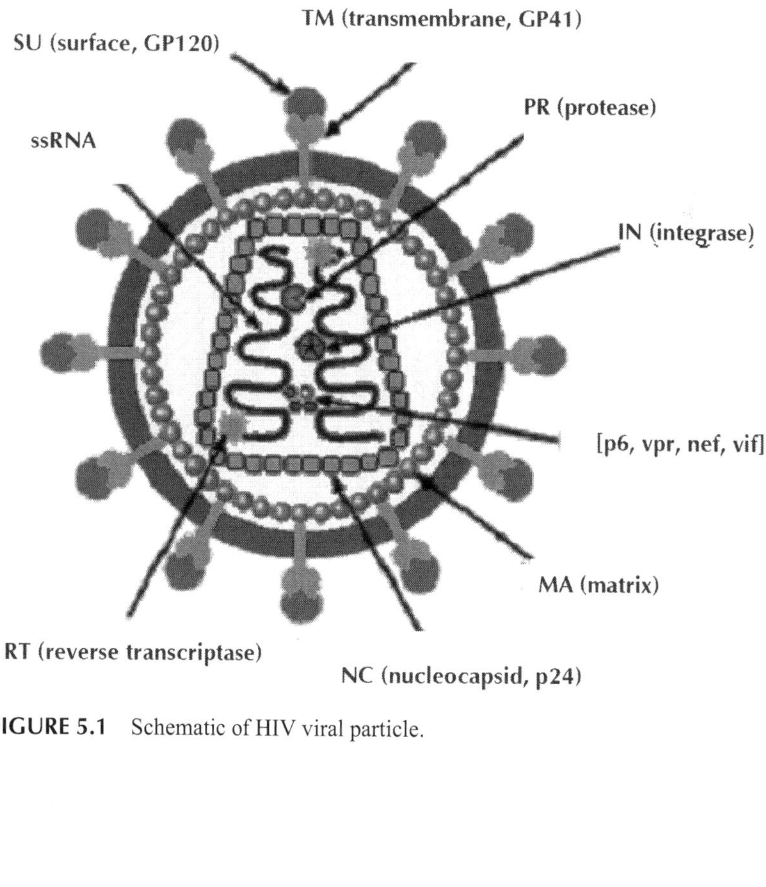

FIGURE 5.1 Schematic of HIV viral particle.

TABLE 5.2 Genes and Proteins of HIV-1

Gene		Gene product and function
gag	**Group-specific antigen**	Core proteins and matrix proteins (p24, p17, and p7).
pol	**Polymerase**	Reverse transcriptase (p66), protease (p11) and integrase (p32) enzymes.
env	**Envelope**	Transmembrane glycoproteins (gp120 and gp 41), gp120 binds CD4 and CCR5; gp41 is required for virus fusion and internalization.
tat	**Transactivator**	Transactivating protein (p14), Positive regulator of transcription.
rev	**Regulator of viral expression**	p19, allows export of unslpiced and partially spliced transcripts from nucleus.
vif	**Viral infectivity factor**	p23, affects particle infectivity.
vpr	**Viral protein R**	p15, Transport of DNA to nucleus, augments virion production, cell cycle arrest.
vpu	**Viral protein U**	p16, Promotes intracellular degradation of CD4 and enhances release of virus from cell membrane.
nef	**Negative-regulator factor**	p27, Augments viral replication *in vivo* and *in vitro*, downregulates CD4, MHC class I and MHC class II.

The RNA genome consists of at least 7 structural landmarks (LTR, TAR, RRE, PE, SLIP, CRS, INS) and nine genes (*gag*, *pol*, *env*, *tat*, *rev*, *nef*, *vif*, *vpr and vpu*) encoding 19 proteins. Three of these genes, *gag*, *pol* and *env*, contain information needed to make the structural proteins for new virus particles. For example, *env* codes for a protein called gp160 that is broken down by a viral enzyme to form gp120 and gp41. The six remaining genes, *tat*, *rev*, *nef*, *vif*, *vpr* and *vpu* (or *vpx* in the case of HIV-2), are regulatory genes for proteins that control the ability of HIV to infect cells, produce new copies of virus (replicate), or cause disease (Abfalterer et al., 2008) (Figure 5.2). Though regulatory genes are not required for the viral replication, viral replication efficiency and infectivity are reduced dramatically in the absence of any of them.

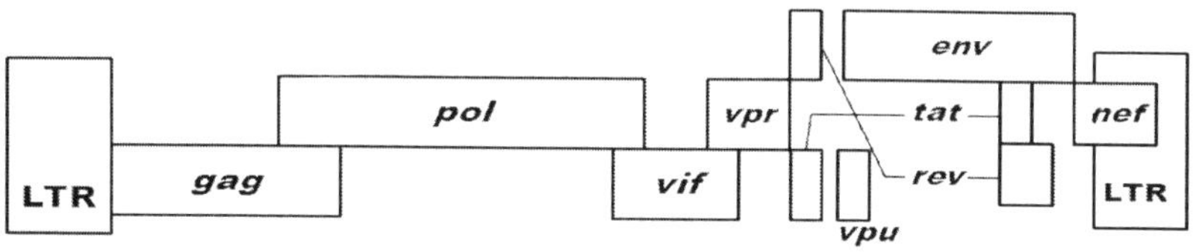

FIGURE 5.2 Organization of the HIV genome.

5.5 AIDS PATHOGENESIS

The defining feature of AIDS, the end stage of HIV-infection, is the rapid decline in the number of CD4+ T lymphocytes (Levy, 1993), the natural host for HIV. During the so-called clinical latency period, which lasts about 10 years, the number or CD4+ cells in the peripheral blood (2% of total) remain steady, and the amount of virus in the blood remains low. Most HIV researchers believe that this steady state results from the direct cytopathic effects of the virus (Wei et al., 1995; Ho et al., 1995) and the continuous production of CD4+ cells and the elimination of infected cells. In the end stage of the disease, the virus wins the protracted battle, and there is a rapid decline in the number of CD4+T lymphocytes concomitantly with acute viremia. Consequently, several body organs become infected with various bacterial and fungal infections followed by reduced immunity. For example, respiratory infections, damage to central nervous system, gastrointestinal disease, candidiasis and kidney disorders etc. are reported as opportunistic infections (OIs). The pathogens mainly include *Pneumocystis jiroveci, Mycobacterium tuberculosis, S. pneumoniae, H. influenza, S. aureus, P.aeruginosa.* HIV-1 is also reported to cause organ specific disease such as HIV-associated nephropathogenesis (Husain et al., 2005; Husain and Singhal, 2011) and HIV-associated neurological disorders (Gelman et al., 2012).

5.6 PREVENTION OF HIV INFECTION

Researchers have shown that several approaches to prevention, when properly executed, can be effective. These approaches include education and behavior modification, the promotion and provision of condoms, the treatment of other sexually transmitted diseases, drug-abuse treatment, access to clean needles and syringes for injection-drug users, and the use of antiretroviral drugs to interrupt the transmission of the virus from mother to infant (Coates and Collins, 1998).

The use of antiretroviral drugs in pregnant women with HIV infection and their infants is an extraordinarily successful prevention strategy (Connor et al., 1994). The rate of mother-to-child transmissions of HIV in the United States has been cut to negligible levels among women and infants treated with an extended regimen of zidovudine therapy. Recent studies by the CDC, the National Institutes of Health (NIH), and others have shown that substantially shorter regimens of anti-retroviral drugs, which would be more feasible in poorer countries, can also reduce perinatal HIV transmission dramatically (Mofenson, 1999; Saba, 1999).

Other methods of preventing HIV transmission may also help slow the epidemic of HIV and AIDS. For example, researchers are developing and testing microbicides, substances that a woman can use in her vagina before sexual intercourse to prevent the transmission of HIV and other Sexually transmitted diseases (Elias and Coggins, 1996).

5.7 ANTIRETROVIRAL THERAPY

Since discovery of HIV, researchers are continuously trying to find a permanent cure of the disease. The complexity of the virus and the frequent mutations occurring in the genome has restricted designing of anti-HIV drugs and development of a novel vaccine that can completely eliminate virus population from the host. But, treatment strategies have been developed through various modes such as administration of anti-HIV drugs, siRNA, herbal medicines, massage therapy, etc. Since treatment administered individually cannot overcome the illness completely, it is believed that all the strategies followed together can contribute significantly.

5.7.1 ANTI-HIV DRUGS

A list of anti-HIV drugs has been approved by Food and Drug Administration (FDA) (Vanessa et al., 2011) that is divided into seven groups as shown in Table 5.3. Additionally, certain drugs are undergoing clinical trials and the list is shown in Table 5.4.

TABLE 5.3 List of Commonly Used Antiviral Compounds (Vanessa et al., 2011)

S. No.	Compound type and generic name	Commercial name	Common Abbreviation	Target
1.	**Nucleoside Reverse Transcriptase Inhibitors (NRTIs)**			
	Azidothymidine/ Zidovudine	Retrovir	AZT	HIV-1 reverse transcriptase
	Zalcitabine	Hivid	ddC	
	Didanosine	Videx	ddI	
	Lamivudine	Zeffix	3TC	
	Stavudine	Zerit	D4T	
	Abacavir	Ziagen	ABC	
	Emtricitabine	Emtriva	FTC	
2.	**Non-Nucleoside Reverse Transcriptase Inhibitors (NNRTIs)**			
	Nevirapine	Viramune	NVP	HIV-1
	Delavirdine	Rescriptor	DLV	reverse transcriptase
	Efavirenz	Sustiva	EFV	
	Etravirine	Intelence	ETV	
	Tenofovir	Viread	TDF	
3.	**Nucleotide Reverse Transcriptase Inhibitors (NtRTIs)**			
	Tenofovir diso-proxil fumarate	Travuda		HIV-1 reverse transcriptase
4.	**Protease Inhibitors (PIs)**			
	Amprenavir	Agenerase	APV	HIV-1
	Indinavir	Crixivan	IDV	protease
	Nelfinavir	Viracept	NFV	
	Atazanavir	Reyataz	ATV	
	Lopinavir	Kaletra / Aluvia	LPV	
	Saquinavir	Invirase / Fortovase	SQV	
	Darunavir	Prezista	DRV	
	Fosamprenavir	Lexiva/ Telzir	FPV	
	Ritonavir	Norvir	RTV	Targets HIV-1 protease and also inhibits cytochrome P450-3A4, which normally metabolizes protease inhibitors

TABLE 5.3 *(Continued)*

S. No.	Compound type and generic name	Commercial name	Common Abbreviation	Target
5.	**Integrase Strand Transfer Inhibitor (INSTI)**			
	Raltegravir	Isentress	RAL	HIV-1 integrase
6.	**Fusion inhibitor (FIs)**			
	Enfuvirtide	Fuzeon	T-20	gp-41 mediated fusion
7.	**Co-Receptor Inhibitor (CRI)**			
	Maraviroc	Selzentry / Celsen-tri	MVC	CCR5 in its role as an HIV-1 Coreceptor

TABLE 5.4 List of Anti-HIV Drugs Currently Undergoing Clinical Trials (Phase II or Phase III) (Youcef and Erik, 2009)

Drug	Class	Developer	Phase
Apricitabine	NRTIs	Avexa	III
Amdoxovir	NRTIs	RFS Pharm	II
Elvucitabine	NRTIs	Achillion Pharmaceuticals	II
Racivir	NRTIs	Pharmasset	II
BILR 355 BS	NNRTIs	Bochringer Ingelheim	II
(+)-calanolide A	NNRTIs	Sarawak MediChem Pharmaceutics	II
IDX899	NNRTIs	Idenix Pharma	I/II
MIV-150	NNRTIs	Medivir, Chiron	II
RDEA806	NNRTIs	Ardea	II
Rilpivrine	NNRTIs	Tibotec	IIb
AK-602	CRIs	Kumamoto University	III
AMD070	CRIs	AnorMed	II
HGS004	CRIs	Human Genome Sciences	II
Ibalizumab (TNX-355)	CRIs	TailMed Biologics	II
PF-232798	CRIs	Pfizer	II
VCH-286	CRIs	ViroChem Pharm	II
Vicrviroc	CRIs	Schering Plough	III
PRO 140	FIs	Progenics	II
SP01A	EI	Samaritan Pharmaceuticals	III
Elvitegravir	INSTIs	Gilead Sciences	III

5.8 COMBINATION THERAPY

A major limiting factor associated with the use of anti-HIV drugs as monotherapy is toxicity and emergence of drug-resistant isolates. To overcome these problems, highly active antiretroviral therapy (HAART) was developed in mid-1990s (Youcef et al., 2009) and was successful in decreasing the morbidity and mortality rates (Correll et al., 1998; Michael et al., 1998; Fauci, 1999; Sendi et al., 1999). In contrast to monotherapy, HAART involves administration of two or more than two drugs, targeting various stages of viral life cycle simultaneously. The major approaches include: (a) two NRTIs or a NRTI with a NNRTI; (b) a NRTI with an antiretroviral agent acting at another step in the viral life cycle; (c) an RT inhibitor with an immunomodulator; and (d) an RT inhibitor with a hematopoietic growth factor. Such formulations are enlisted in Table 5.5. The administration of these formulations has increased adherence to medication, lowered drug resistance and toxicity and ultimately has improved the quality of lives of HIV-positive individuals.

TABLE 5.5 List of the Currently FDA-Approved Anti-HIV Drug Combinations (Youcef and Erik, 2009)

Combination	Components	Manufacturer
Combivir	Zidovudine (300 mg), Lamivudine (150 mg)	GlaxoSmithKline
Trizivir	Abacavir (300 mg), Zidovudine (300 mg), Lamivudine (150 mg)	GlaxoSmithKline
Epzicom (U.S.) / Kivexa (Europe)	Abacavir (600 mg), Lamivudine (300 mg)	GlaxoSmithKline
Truvada	Tenofovir disoproxil fumarate (300 mg), Emtricitabine (200 mg)	Gilead Sciences
Atripla	Emtricitabine (200 mg), TDF (300 mg), Efavirenz (600 mg)	Bristol-Myers Squibb, Gilead Sciences

5.8.1 RECEPTOR BINDING AND FUSION INHIBITORS

5.8.1.1 RCD4 AND CONGENERS

CD4, is a cell surface glycoprotein and a primary receptor for the attachment of HIV to human cells. The attachment is through high-affinity binding with the HIV envelope glycoprotein gp120. Truncated, soluble molecules, which contain the extracellular, N-terminal domain of CD4 have been produced by using recombinant DNA technology and these recombinant soluble CD4 (rCD4) molecules maintain their ability to avidly bind HIV gp120 (Smith et al., 1987). rCD4 can be used as antiviral agent as it can be complexed with toxic substances and can target HIV infected cells expressing gp120 on their cell surfaces. For example, hybrid molecules have been produced by linking rCD4 domains to toxins such as *Pseudomonas* exotoxcin or ricin. The CD4 portion of the hybrid selectively binds to HIV infected Cells and the toxin kills the cell. A major limitation of the above strategy is the expression of gp120 on HIV infected cells. In this regard, it is believed that *Pseudomonas* exotoxin can be synergized with AZT so that the current antiviral strategy can act in a better way. It is studied that exposing HIV to an excess of rCD4 might block the virus's ability to bind to CD4, due to competitive inhibition. Also, it is seen that binding of rCD4 to gp120 may stip away gp120 from the virion (Moore et al., 1990). This can interfere with binding to CD4 cells or directly damage the virion. Thus, rCD4 has shown to be effective in blocking acute infection (Smith et al., 1987) of lymphoid cells and moncyte/macrophages (Harbison et al., 1990). Various modifications of rCD4 such as hybrid, CD4-Ig are underway for clinical trials. The hybrid has been reported to mediate antibody-dependent cell-mediated immunity against HIV-infected cells (Byrn et al., 1990). Simultaneously, certain problems are worth noting that limit the use of rCD4. These include: (i) interference with normal CD4 cell function; (ii) formation of antibody against rCD4 protein. The gp41, which is a transmembrane subunit of the HIV envelope protein, is a vulnerable target to inhibit HIV entry. The synthetic fusion inhibitorT20 (brand name: Fuzeon, generic name: enfuvirtide) available by prescription. However, it has several drawbacks such as a high level of development of drug resistance, a short-half life *in vivo*, rapid renal clearance and low oral bioavailability (Yi et al., 2016).

5.8.1.2 DEXTRAN SULFATE AND OTHER SULFATED POLYANIONS

A group of sulfated polysaccharide, polyanionic compounds is known to block the binding of HIV to its receptor, CD4 (Baba et al., 1990). These include dextran sulfate, heparin and pentosan. It is reported that the antiviral effect of dextran sulfate and heparin is due to interaction with positively charged amino acids in the V3 loop region of gp120 (the principal neutralizing domain) (Callahan et al., 1991).

5.8.1.3 PLANT LECTINS

Another class of compounds that are D-mannose-specific plant lectins, interfere with an early step of binding to host cells. The mechanism of action is not known and hence, further studies are underway.

5.8.2 TRANSCRIPTION AND TRANSLATION INHIBITION

The HIV *pol* gene encodes a number of enzymatic functions related to reverse transcription. First, a DNA polymerase (the RT) assembles a DNA strand complementary to the viral RNA. Then the RNA strand must be removed to allow a complementary DNA strand to form double-stranded DNA. The C-terminal region of the HIV RT enzyme contains a ribonuclease, RNase H, which serves this function. Mutations in the RNase H domain of the RT have been shown to abolish infectivity. Researchers are looking forward into the development of inhibitors of RNase H (Yu et al., 2008). Certain natural compounds such as β-thujaplicinol and manicol, derived from heartwood of cupressaceous trees have been identified with anti-HIV activity affecting Ribonuclease H at sub-micromolar concentrations (Budihas et al., 2005).

5.8.2.1 ANTISENSE OLIGONUCLEOTIDES

Oligonucleotides which are complementary to a specific segment of a viral RNA genome or mRNA transcript can be synthesized, and by hybridizing with that segment, these oligonucleotides interfere with its normal expression or function. These anti-sense oligonucleotides are directed against viral replication by interference with transcription of specific regulatory or structural genes and/or translation and splicing of mRNA. Initially, anti-sense oligonucleotides were reported with limited therapeutic potential because of problems with poor permeability into target cells and poor stability because of degradation by host nucleases. Subsequently, phosphate-modified oligonucleotides were developed as they were found to have improved stability and permeability. These synthetic molecules were found to be effective against HIV-1 and HIV-2 in *in vitro* conditions in the millimolar range (Agarwal et al., 1989; Matsukura et al., 1987, 1989; Stein et al., 1989). Such molecules have been designed against segments of the *rev* and *gag* genes. A methylphosphonate antisense oligonucleotide directed against an enhancer element in the U3 region has been shown to inhibit induction of chronic HIV infection *in vitro* (Laurence et al., 1991).

5.8.2.2 RIBOZYMES

The use of ribozymes is another rapidly growing strategy that can be used as antiviral therapy. Ribozymes are known to possess enzymatic, self-cleaving activity. A number of ribozymes have been isolated from plant RNA pathogens and a family of ribozymes with a "hammerhead" conformational structure has been shown to recognize specific sites on HIV RNA and cleave viral RNA. The use of such molecules is limited to a greater extent because of toxicity to host RNA, problem of delivery and high mutation rate of HIV (Sarver et al., 1990).

5.8.2.3 TAT, REV AND NEF INHIBITORS

Apart from the seven major targets, discussed above, recently novel targets have been identified (Youcef et al., 2009) that include regulatory genes of HIV-1. It is believed that since TAT, REV and NEF are critical for viral survival, targeting such proteins might potentiate the fight against HIV-1 (Yu et al., 2008). Strategies to attack these regulatory elements include: the use of anti-sense oligonucleotides, ribozymes, ineffective mimics of the regulatory proteins which bind to and block the responsive regulatory site but do not cause activation, inhibitors of protein processing which block the splicing of the regulatory gene products, monoclonal antibodies or vaccines directed against the soluble Tat protein, or other compounds which inhibit expression of these regulatory genes or their protein products. Inhibitors of tat or the Tat product are in development. One such candidate compound, RO 24-279, a benzodiazepine derivative, appears promising and has entered a phase I trial, but the future developmental direction for this agent is uncertain (Palca, 1991).

It is very much clear that NEF plays critical role in AIDS pathogenesis. The *nef* is known to prevent the death of infected cells, contributing to increased viral load. The mechanism involves prevention of apoptosis of infected cells by either inhibiting proteins involved in apoptosis or preventing the infected cells from being recognized by Cytotoxic T Lymphocytes. Neutralization of *nef* can hasten the death of infected cells and help reduce the viral load. *Nef* is therefore a very important molecular target for developing therapeutics that slows disease progression to AIDS. The N-terminal region of *nef* and the naturally occurring bee venom mellitin have very similar primary and tertiary structures, and they both act by destroying membranes. Chemical analogs of a mellitin inhibitor prevent *nef*-mediated cell death and inhibit the interaction of NEF with cellular proteins involved in apoptosis. Naturally occurring bee propolis also contains substances that prevent *nef*-mediated cell lysis and increases proliferation of CD4 cells in HIV-infected cultures. These chemical compounds and natural products are water soluble and nontoxic and are therefore can be used as potential drugs (Ahmed, 2005).

^{1}G**GK**W**SK**SSV**IGWPA**VRE**RMRR**^{21} **Nef**
GI**GA**V**LK**VLT**TGLPA**LIS**WIKR** **Mellitin**

Sequence similarity between N-terminal region of Nef and Mellitin.

5.8.3 MATURATION AND RELEASE INHIBITORS

5.8.3.1 IFN-α

Shortly after the discovery of HIV, alpha interferon was shown to inhibit the replication of HIV *in vitro* (Ho et al., 1985). Intravenous administration of IFN-α has also showed moderate benefit in delaying disease progression (Lane et al., 1990). Also, IFN-α administration was found to be associated with significant toxicity, including granulocytopenia, elevated liver enzymes, and flulike symptoms. Therefore, role of IFN-α as a single antiretroviral agent is limited, but studies employing IFN-α as combination regimens are ongoing.

5.8.3.2 PROTEASE INHIBITORS

The HIV *pol* gene encodes a protease enzyme which is required for post-translational cleavage of a precursor p55 Gag protein into four structural core proteins, and for cleavage of the p160 Gag-Pol precursor into RT, endonuclease, structural proteins and the HIV protease itself (Jacks et al., 1988). A series of HIV inhibitors has been designed on the principle of the transition-state mimetic concept (Roberts et al., 1990). Synthetic peptido-mimetic compounds which bind to the protease but whose scissile amide bonds are replaced by nonhydrolyzable elements, such as phosphinic acid, reduced amide, hydroxyethylene, or hydroxyethylamine have been synthesized. Such compounds have shown potent anti-HIV activity in the nanomolar range *in vitro*, without cytotoxicity at effective levels (Ashorn et al., 1990; Dreyer et al., 1989; McQuade et al., 1990; Meek et al., 1990; Roberts et al., 1990). These compounds are reported with certain side effects such as cytotoxicity, poor oral bioavailability because of digestion or binding to gut proteases and rapid metabolism. To overcome these problems, a protease inhibitor was designed that had reduced peptidic characteristics with potent, specific anti-HIV activity and improved stability to degradative enzymes (Erickson et al., 1990). A representative of this class of compounds is under clinical trials in Europe. Similarly, VIF is a protease belonging to cysteine family and cysteine protease inhibitors can be used to target the protein.

5.8.3.3 GLYCOSYLATION INHIBITORS

Cellular enzymes are known to participate in posttranslational modification of viral proteins through processes such as glycosylation and myristoylation. Glucosidase inhibitors such as castanospermine and N-Butyl-deoxynojirimycin have been developed and shown to block HIV infectivity *in vitro* and to block syncytium formation (Karpas et al., 1989; Ratner et al., 1991). Also, N-Butyl-deoxynojirimycin has entered phase I trials.

5.8.3.4 MYRISTOYLATION INHIBITORS

Myristoylation inhibitors which block the addition of myristic acid to HIV Gag proteins have been identified. A number of these agents were shown to decrease productive HIV infection *in vitro*, but clinical development of these agents was slowed by significant cellular toxicity. Recently, a series of myristate analogs in which a methylene group is replaced by a sulphur or oxygen atom has been shown to inhibit acute and chronic HIV-1 infection in H9 cells without cellular toxicity at effective drug levels (Bryant et al., 1991).

5.8.3.5 HYPERICIN

Hypericin and Pseudohypericin are aromatic polycyclic diones isolated fom the plant family Hypericum (St. John's wort). Hypericin is thought to impair the assembly and release of mature HIV virions (Meruelo et al., 1988). The mechanism is unknown; hypericin may have direct effects on the HIV virion or may interact with the budding virus at the cell membrane. It has also been reported to be an inhibitor of protein Kinase C (Takahashi et al., 1989). Human trials of herbal extracts containing hypericin have not shown a clear benefit. A phase I trial of an intravenously administered synthetic hypericin compound is underway.

5.8.4 IMMUNOMODULATION

A major parallel direction in the field of antiretroviral therapy is the development of agents or techniques, which will enhance or restore the host's immune system. The strategies employed here include the use of chemical stimulants of the immune system, cellular or humoral immune reconstitution, and vaccination.

5.8.4.1 CHEMICAL IMMUNE STIMULANTS

Ampligen is a mismatched, double-stranded RNA molecule [poly(I): poly(C12-U)*n*] which may enhance immune function through a variety of postulated mechanisms including promotion of IFN-like antiviral activity (Carter et al., 1987). It may also have direct anti-HIV activity acting at an early stage in viral replication (Montefiori et al., 1989). Initial clinical trials of intravenously administered ampligen suggested beneficial effects

(Carter et al., 1987; Strayer et al., 1990), but doubt remains concerning this agent's true efficacy.

Diethyldithiocarbamate (dithiocarb or DTC) is a drug which has been reported to have beneficial effects in patients with HIV infection but whose mechanism of action is unclear. It possesses antioxidant and chelating activity, inhibits prostaglandin E2 production, and appears to affect T-cell differentiation (Hersh et al., 1991; Reisinger et al., 1990). Its effects on HIV replication *in vitro* are debated, but DTC probably has little direct antiviral effect.

Thymic humoral factor and thymopentin, a synthetic peptide mimicking the active site of thymopoietin, are under study. Preliminary results of two recent placebo-controlled studies of thymopentin involving a total of 136 asymptomatic or mildly symptomatic patients (Conant et al., 1990; Thompson et al., 1990) have indicated benefit in maintaining CD4 lymphocyte counts and possibly in slowing disease progression.

5.8.3.2 VACCINATION OF HIV POSITIVE INDIVIDUALS

The best long-term hope for controlling the AIDS pandemic is a safe, effective and affordable preventive vaccine, but its development has encountered unprecedented scientific challenges. The first phase trial of an HIV vaccine was conducted in 1987. Subsequently, more than 30 candidate vaccines have been tested in over 60 phase I/II trials, involving approximately 10,000 healthy volunteers. Most of these trials have been conducted in USA and Europe, but several have also been conducted in developing countries. The first phase III trials began in the USA in 1998 and in Thailand in 1999 to assess the efficacy of the first generation of HIV vaccines (based on the HIV envelope protein, gp 120) (Jose, 2001). To accelerate the development of an HIV vaccine, additional candidate vaccines must be evaluated in parallel in both developed and developing countries. This essentially requires international collaboration and coordination. For the future prevention of AIDS, efforts must be made in the direction of best use of vaccines.

5.9 CHALLENGES

The Scientific community has been facing several problems in the development of vaccine candidates. Three are majorly considered responsible, that is, (a) lack of information on the immunological correlates of protection against HIV/AIDS; (b) genetic variability of HIV; and (b) lack of good animal models.

5.10 CANDIDATE VACCINES

Based on the nature of vaccine, these are classified as first-generation, second-generation and third-generation vaccines. The first-generation vaccines were based on the envelope proteins of HIV, especially gp120. Second-generation vaccines are designed to induce cell-mediated immunity, using either the live vectors (such as vaccinia, canarypox, etc.) or "naked" DNA that codes for different HIV genes. Third-generation vaccines, based on regulatory proteins of HIV, such as *tat* and *nef* are also emerging (Jose, 2001).

In 1999, NIAID (National Institute of Allergy and Infectious diseases) planned to establish an integrated network called, HVTN (HIV Vaccine Trials Network). The HVTN is an international collaboration of scientists and educators facilitating the development of HIV/AIDS preventive vaccines. The HVTN conducts all phases of clinical trials, from evaluating experimental vaccines for safety and immunogenicity, to testing vaccine efficacy and improving the process of vaccine designing (James et al., 2012).

The HVTN successful completion of so many HIV vaccine clinical trials has provided a wealth of information on both safety and immunogenicity of a large number of diverse products and vaccine regimens, from single protein and DNA products to prime boost combinations (Table 5.6). In collaboration with Merck, the HVTN participated in the STEP study, which was the first to examine efficacy of a vaccine designed to elicit strong T-cell responses and was also the first to use a phase IIb design (Buchbinder et al., 2008). A major advance in the field has been the co-administration of DNA vaccines and cytokines through electroporation, resulting in dramatic enhancements of the CD4+ and CD8+ T-cell immune responses (Table 5.7).

TABLE 5.6 HIV Vaccine Trials Network Trials By Year (James et al., 2012)

	2000	**2001**	**2002**	**2003**	**2004**	**2005**	**2006**	**2007**	**2008**	**2009**	**2010**	**2011**	**Total**
Trials opened	1	2	2	6	7	5	4	5	1	11	3	10	57
Participants enrolled	14	389	282	338	1336	1180	2247	1216	124	3614	1616	1251	13,607
Publications	12	10	5	13	15	16	14	14	18	20	27	28	192
Trials by product													
Protein			1		2								3
DNA				2	2	1		1		1			7
Viral Vector		1		2	3	2		3		2	1	4	18
DNA and Protein				1		1	1					1	4
DNA and Viral Vector	1					1	3	1		3	1	2	12
Viral Vector and protein		1										1	2
Trials by phase													
I		2	1	5	6	3	2	2		2		5	28
Ib					1		2	2		2	2	3	12
IIa	1					1				1			3
IIb						1		1		1			3
Other			1	1					1	5	1	2	11

TABLE 5.7 Immune Responses Generated By Different Vaccine Platforms (James et al., 2012)

Vaccine Type	CD4$^+$ T cells	CD8$^+$ T cells	Binding antibody	Neutralizing antibody
Naked DNA	++	+/-	++	+/-
Cytokine adjuvanted DNA*	+/-	+/-	-	-
Electroporated DNA*	++	++	++	+
Pox vectors	++	+	++	+
Adenovirus vectors	+	+++	++	+
gp120 protein	-	-	++	+

+/– denote relative strengths of the immune response, with more + indicating a stronger response.

*Denotes additional effects on top of naked DNA.

Another recent focus that has been on a series of small Phase Ib studies that aim to address basic science questions and generate new hypotheses regarding vaccination strategies and their associated immune responses. Vaccine candidates that progress into efficacy trials have the greatest potential to advance the field and generate immune correlates of protection. The largest ongoing HIV vaccine trial, known as HVTN 505, is a Phase IIb trial testing a vaccine regimen developed by the Vaccine Research Center at NIAID. The regimen, a DNA prime followed by a recombinant adenovirus type 5 (Ad5) boost, was found to be the most immunogenic in any HVTN study to date. The results of this trial, together with the large amount of immunogenicity data being collected, are expected to provide valuable information for future vaccine developments.

Improving on the efficacy and durability of the RV144 vaccine regimen is a primary focus of current HVTN activities. Recognizing both the immense need and challenge of developing an efficacious HIV vaccine, a novel collaboration has been created between pharmaceutical companies and nonprofit organization, known as the Pox Protein Public Private Partnership (P5). The primary aim of the collaboration is to extend and confirm the RV144 findings in other geographical locations, such as South Africa, and to prepare a path to eventual vaccine licensure.

Completion and follow-up analysis of the numerous ongoing HVTN clinical trials, and the HVTN 505 Phase IIb trial in particular, are expected to provide valuable insight into future vaccine development strategies.

These and future trials also provide a valuable opportunity to perform research on behavioral aspects of clinical trial participation (James et al., 2012).

5.11 HIV-1 ENVELOPE (*ENV*)

Since *env* is the only viral protein on the outside of the HIV-1 virion, neutralizing antibodies can be raised against gp120 and gp41. Thus, *env* is of vital importance for vaccine research. Studies focusing designing of *env*-based vaccines, from the angle of structural biology and development of therapeutic antibodies are undergoing (Pejchal and Wilson, 2010; Abele et al., 2010).

5.12 ATTENUATED POXVIRUS VECTORS: MVA AND NYVAC

Two of the most promising poxvirus vectors are the highly attenuated modified vaccinia virus Ankara (MVA) and the modified Copenhagen strain NYVAC. MVA was obtained from chorioallantoid vaccinia virus Ankara (CVA) after 570 passages by Mayr and colleagues. The attenuated mutant developed in such a manner lost about 30 kb, at both ends of the viral genome and was termed MVA. Similarly, using the Copenhagen vaccine strain of vaccinia virus and gene targeting techniques, Tartaglia and colleagues produced a virus termed as NYVAC, with 18 viral genes inactivated (Tartaglia et al., 1992). These vectors have shown a good safety profile and are known to trigger strong, broad and durable immune responses to HIV antigens in humans. NYVAC has been shown to drive the immune response towards CD4+T cells, whereas MVA towards activation of CD8+T cells. Both vectors have biological properties such as induction of cytokines and chemokines, intracellular signaling pathways and nature of poly-functional responses (Mariano, 2009). It is also found that MVA vectors with viral genes deleted antagonize host specific immune responses and ultimately benefit immune system (Staib, 2005).

Since, the two vectors fulfill some of the functions of a novel vaccine, it is believed that the basic understanding of the molecular interaction between virus vectors and host together with their immunological behavior will provide a rationale for optimal use of MVA and NYVAC vectors as HIV vaccines.

5.13 MICROBICIDES

Worldwide, nearly half of all individuals living with HIV are women, who acquire the virus largely by heterosexual exposure. With an HIV vaccine likely to be years away, topical microbicide formulations applied vaginally or rectally are being investigated as another strategy for HIV prevention.

Microbicides are products that are designed for application at vaginal or rectal mucosae to inhibit or block early events in HIV infection and thereby prevent transmission of HIV. Currently, the most advanced microbicides in the development pipeline are based on highly active anti-retroviral drugs (ARVs) (Kelly and Shattock, 2011).

The sexual transmission of HIV is not uniformly efficient. The type of sexual activity and the phase of disease affect the risk of transmission. Recent investigation has shown that the rate of sexual transmission depends on cofactors such as circumcision status, genital ulcer disease, and phase of disease (Powers et al., 2008). High serum HIV-1 concentrations during the acute infection period increase the probability of male-to-female heterosexual transmission by upto eight to tenfold (Pilcher CD et al., 2004). Male circumcision status also affects the efficiency of transmission. It decreases the risk of female-to-male HIV transmission by 50–76% (Auvert et al., 2005; Gray et al., 2007).

Topical microbicides are grouped into five classes of agents, based on where they disrupt the pathway of sexual transmission of HIV. These classes include surfactants/membrane disruptors, vaginal milieu protectors, viral entry inhibitors, reverse transcriptase inhibitors, and a fifth group whose mechanism is unknown. A detailed list of microbicides developed is given in Tables 5.8 and 5.9 (Blayne and Jessica, 2008).

TABLE 5.8 Non-Specific Microbicide Agents (Blayne and Jessica, 2009)

	Advantages	Disadvantages	Examples in class
Surfactants			
Non-specific disruption of cellular and microbicidal membranes	Active against wide range of pathogens; often spermicidal	Potentially toxic to host cells	Nonoxinol 9, C31G (Savy), Sodium lauryl sulfate
Vaginal milieu protectors/acidifying agents			
Restores protective acidic pH of vagina by buffering semen	Spermicidal activity against HIV, HSV, *C. trachomatis*	None known	Carbopol 974P (Buffer gel), Acid-form (Amphora)

TABLE 5.8 *(Continued)*

	Advantages	Disadvantages	Examples in class
Entry inhibitors: anionic polymers			
Negative charge causes interaction with HIV's viral envelope proteins and interferes with attachment of HIV to CD4+ cells	Many have activity against other STI pathogens (including *C. trachomatis, Neisseria gonorrhea, and HSV)*	Not all virus types respond equally well to negative charge properties of these compounds	Napthalene sulfonate (PRO2005), Carrageenan (Carraguard/PC-515), Cellulose sulfate (Ushercell), Cellulose acetate phthalate (CAP), Dendrimers: SPL7013(Vivagel)

TABLE 5.8 Specific Microbicide Agents (Blayne and Jessica, 2009)

	Advantages	Disadvantages	Examples in class
Entry inhibitors: CCR5 blockers			
Block CCR5 co-receptor and interfere with attachment of HIV to host cells	Targets specific ligands	No activity against other STI pathogens	PSC-RANTES CMPD167
Reverse transcriptase inhibitors			
Interfere with HIV reverse transcriptase enzyme	Tenofovir: active in multiple cell types. TMC-120 and UC781 (NNRTIs): delayed development of resistance compared with first-generation NNRTIs	No activity against other STI pathogens	Tenofovir (PMPA; nucleotide analogue), TMC120 (NNRTI), UC781 (NNRTI)

5.14 SIRNA

Small interfering RNA (siRNA) is small RNA of 18-25 nucleotides (nt) in length that play important role in regulating gene expression. It is incorporated into an RNA-induced silencing complex (RISC) and serves as guides for silencing their corresponding target mRNAs based on complementary base-pairing. The promise of gene silencing has led many researchers to consider siRNA as an anti-viral load tool. However, in long-term settings, many viruses appear to escape from this therapeutical strategy. HIV sets a very good example in evading RNA silencing, by either mutating the siRNA-targeted sequence or by encoding for a partial suppressor of RNAi (RNA interference) (Man et al., 2005).

5.15 STEM CELL BASED APPROACHES

A primary therapeutic goal towards treating HIV infection is the enhancement of natural immune responses that can lead to the direct clearance of the virus from the body. Therefore, use of Haematopoeitic Stem Cells (HSCs) represents a potentially powerful approach that can help restoration of total functional immunity to the affected individual. The HSC approach involves reconstitution of immune responses that can overcome the barriers necessary to clear the virus from the body and protection of newly developed cells from HIV infection, thus preventing them from becoming another reservoir of infected cells. Coupled with the currently available anti-retroviral drugs and other therapeutic strategies, the HSC approach can become a viable HIV treatment strategy (Scott and Jerome, 2011).

5.16 DENDRITIC CELLS

During sexual transmission of HIV, virus crosses mucosal epithelium and eventually reaches lymphoid tissue where it establishes a permanent infection. Evidence has accumulated that infection of Langerhan cells, which are resident dendritic cells in pluristratified epithelia, plays a critical role in the early events in HIV transmission. HIV infection of Langerhan cells is regulated by surface expression of CD4 and CCR5. Thus, topical microbicides that interfere with HIV infection of Langerhan cells represent an attractive strategy for blocking sexual transmission of virus. Capture and uptake of HIV virions is another major pathway by which HIV interacts with dendritic cells. This process is mediated by a C-type lectin, DC-SIGN. It is known that dendritic cell-T cell interaction, critical in the generation of immune responses, forms a rich microenvironment for HIV replication. Dendritic cells that have captured virus via DC-SIGN, facilitates infection of T cells in chronically infected individuals. Therefore, blocking DC-SIGN-mediated capture of HIV represents a potential therapeutic antiviral strategy for HIV disease. Also, dendritic cells have been targeted both *ex vivo* and *in vivo* to initiate and enhance HIV-specific immunity (Vincent and Andrew, 2002). Injection of DNA coated particles via gene guns and application of protein patches represent certain vaccine strategies aiming at exploiting the immune-stimulating potential of Langerhan cells within skin (Johnston and Tang 1994; Glenn et al., 2000).

A vaccine candidate targeting dendritic cells in the lymph nodes and other organs of the immune system is under trial. The vaccine candidate contains a monoclonal antibody (mAb) engineered to recognize DEC-205, an endocytic protein found on the surface of dendritic cells that mediates efficient presentation of antigens. The mAb was fused to an HIV clade B p24 Gag protein. Gag p24 was selected as the vaccine antigen because of its many conserved epitopes, which can induce greater breadth of CD8+ T-cell responses. The vaccine candidate was administered in 45 healthy HIV-uninfected individuals with an adjuvant, Poly ICLC (Hiltonol). The adjuvant was used to activate innate immune responses or to help immature dendritic cells. The study was conducted in Rockefeller University in New York (McEnery, 2010).

5.17 TRADITIONAL INDIAN MEDICINE AND HOMEOPATHY FOR HIV/AIDS

The Indian health system has perhaps the world's largest community-based indigenous system of medicine, and it includes Ayurveda, Unani medicine, Siddha medicine, yoga and naturopathy (http://mohfw.nic.in/reports/Annual0506/Ayush%20annual%20report%20final.pdf). These forms of traditional Indian medicine along with homeopathy (termed as TIMH) are commonly used. Reasons for TIMH use include a strong belief in TIMH efficacy as a "natural" and holistic option, and the fact that allopathic care is often costly, inaccessible and culturally dissonant. While the exact number of non-allopathic providers is not known as many are unregistered. India's Ministry of Health and Family Welfare has reported that there are over 700,000 registered practitioners of TIMH (Financing and Delivery of Health Care services in India, 2005). Over 65% of the population in rural areas of India is using TIMH and medicinal plants to help meet their primary health care needs (World Health Organization, Traditional Medicine: Report by the Secretariat, 2003). TIMH is used to treat a wide variety of conditions, including cancer, diabetes and HIV/AIDS (Aggarwal, 2006; Banerji, 1981; Deivanayagam et al., 2001; Singh et al., 2005). Little is known about TIMH use or its risk and benefits in HIV/AIDS management (WHO: Traditional Medicine strategy: 2002–2005). A survey of 1667 HIV-infected persons in 4 regions of India found that 41% reported using some form of TIMH although only 5% believed TIMH was more effective than allopathic ART (Ramachandani et al., 2007). Studies that reported effectiveness and improved outcomes include those of Ayurvedic and homeopathic treatments such as Boxwood (*Boxus sempervirens*), Andrographolide (*Andrographis paniculata*), and neem (*Azadirachta indica*) as well as the Siddha combination therapy RAN (*Rasagandhi mezhuga, Amukkara chooranum* and *Nellikkai lehyam*) (Deivanayagam et al., 2001). With many products prepared locally as well as available in the market and claims of "cure" being made (Klein, 2007), there is a need for patients, providers and policy makers to assess systematically the potential benefit as well as potential harm associated with TIMH therapies for HIV/AIDS. In light of the suboptimal effectiveness of vaccines, ART, barrier methods and behavior change strategies for prevention and cure of HIV infection, it is both important and urgent to develop a collaborative research agenda that uses rigorous methodologies to investigate, evaluate and better understand the role of TIMH in managing HIV/AIDS and associated illnesses in India.

5.18 MASSAGE THERAPY

In a study conducted by Susan, massage therapy has been proposed to have positive effect on quality of life and on immune function through stress mediation in people living with HIV/AIDS (2010). In particular, HIV/AIDS is rife in developing and low-income countries where health budgets are small. Massage has been identified as a suitable, low cost intervention in the context of developing countries (Maulik, 2009). Massage therapy is defined as systematic and scientific manipulation of body tissues performed with the hands of the therapist for the purpose of affecting the nervous and muscular systems and the general circulation, and is administered by a qualified health professional (Susan et al., 2010). In the study by Susan, it was proved that massage therapy in combination with other modalities, such as biofeedback stress reduction (Birk, 2000) and meditation (Williams, 2005) was superior to massage therapy alone or the other modalities alone (2010) . Also, immunological functions were found to be improved. For example, CD4 count, markers of natural killer cells were increased (Diego, 2001; Shor-Posner, 2006). Diego also reported that massage therapy has shown positive effect in reducing depression. It was found that the effects of massage therapy on relaxation responses are short lived and that combining it with other methods that promote long-term stress management offers more benefits.

5.18.1 AWARENESS, SOCIAL STIGMA AND SOCIOECONOMIC CONSEQUENCES

HIV/AIDS awareness programs are frequently delivered to curb the problem. Despite several efforts, the awareness is limited to urban areas only. A major proportion of AIDS patients are the rural population that has been neglected. At many VCTC centers, cartoon-based educational materials developed for low-literacy populations to convey, simple, comprehensive messages on HIV transmission, prevention, support and care are generally found. Simultaneously, street theatre cultural programs can be used to highlight issues related to AIDS and stigma. Low awareness and high stigma regarding HIV and sex/sexuality-related issues is fuelled by socio-economic conditions of poverty, low literacy and cultural traditions that consider sexual topics taboo (Van, 2004). Spousal communication about sex and sexual health is limited. Due to gender inequity, women have little or no ability to negotiate safe sex and are left vulnerable to infection, violence and stigma (Go et al., 2003; Sivaram et al., 2005). Rural women are especially vulnerable to infection, as many of them are trapped in sociocultural conditions of subordination, are confined largely to their village and immediate surroundings, and are denied access to information, medical treatment, or the ability to protect themselves against potentially unsafe sex with their husband (Koen et al., 2008). Mass media campaigns (such as radio, television, and posters) focus usually on a limited spectrum of messages about sex and condom use. More comprehensive sources of HIV information (such as brochures) are often available at the larger distinct hospitals but usually do not reach the healthcare facilities at the village level. Additionally, in the absence of a trained educator or counsellor who has time to provide a complete explanation many people are shy or afraid to ask HIV- or sex-related questions, and such information does not reach people with low literacy (Chatterjee et al., 2004; Kattumuri et al., 2003). It is very much essential to look forward in the direction of curbing AIDS-related stigma and ultimately discrimination. The social stigma has engulfed the people to such an extent that HIV infected patients do not feel free to share their agony. Stronger steps must be taken such as comfortable communication through counsellinb, maintaining privacy and providing care and support to the sufferer to enhance their mental and social wellbeing.

In India, the socioeconomic consequences of AIDS include loss of job, selling of assets, school dropouts, decreased family income, and increased expenditure. The well-studied reasons include illness, disclosure of sero-status, expense of own sickness, expenses for spouse's sickness, illness related to HIV infection of child, financial problems, death of a parent, own sickness and unemployment, spouse's sickness, medicine, investigation, travel, hospitalization, etc. It is observed that the family relationships are largely affected. Women faces strained relationship with in laws and are blamed of spouse's Illness also (Tarapdhar et al., 2011).

National and international agencies should dedicate more funding to expand and replicate peer education models in rural areas that are in urgent need of similar activities to avert an increase in HIV prevalence.

KEYWORDS

- **AIDS**
- **antiretroviral therapy**
- **HIV vaccine**
- **HIV-1**
- **microbicide**
- **Nef**
- **negative regulatory factor**

REFERENCES

Abele, I. A., Reynell, L., Trkola, A. Therapeutic antibodies in HIV treatment-classical approaches to novel advances. Curr. Pharm. Des. 2010, 16(33), 3754–3766.

Abfalterer, W., Gayathri, A., Will, F., Bob, F., Brian, G., Peter, H., Chien-Chi, Lo.; Jennifer, M., James, J. S., James, T., Hyejin, Y., Ming, Z. HIV Sequence Compendium. 2008.

Aggarwal, B. From traditional Ayurvedic medicine to modern medicine: identification of therapeutic targets for suppression of inflammation and cancer. Expert. Opin. Ther. Targets. 2006, 10(1), 87–118.

Agrawal, S., Ikeuchi, T., Sun, D., Sarin, P. S., Konopka, A., Maizel, J., Zamecnik, P. C. Inhibition of human immunodeficiency virus in early infected and chronically infected cells by antisense oligodeoxynucleotides and their phosphorothioate analogues. Proc. Natl. Acad. Sci. USA. 1989, 86(20), 7790–7794.

Ahmed, A. A. novel drugs and vaccines based on the Structure and function of HIV Pathogenic Proteins Including Nef. Annals New York Academy of Sciences. 2005, 279–292.

Ashorn, P., McQuade, T. J., Thaisrivongs, S., Tomasselli, A. G., Tarpley, W. G., Moss, B. An inhibitor of the protease blocks maturation of human and simian immunodeficiency viruses and spread of infection. Proc. Natl. Acad. Sci. USA. 1990, 87(19), 7472–7476.

Auvert, B., Taljaard, D., Lagarde, E., Sobngwi-Tambekou, J., Sitta, R., Puren, A. Randomixed, controlled intervention trial of male circumcision for reduction of HIV infection risk: the ANRS 1265 Trial. PLos Med. 2005, 2, e298.

Baba, M., Schols, D., Pauwels, R., Nakashima, H., De Clercq, E. Sulfated polysaccharides as potent inhibitors of HIV-induced syncytium formation: a new strategy towards AIDS chemotherapy. J. Acquir. Immune. Defic. Syndr. 1990, 3(5), 493–499.

Balzarini, J. Targeting the glycans of glycoproteins: a novel paradigm for antiviral therapy. Nat. Rev. Microbiol. 2007, 5, 583–597.

Balzarini, J., Van, L. K., Daelemans, D., Hatse, S., Bugatti, A., Rusnati, M., Igarashi, Y., Oki, T., Schols, D. Pradimicin A, a carbohydrate binding nonpeptide lead compound for treatment of infections with viruses with highly glycosylated envelopes, such as human immunodeficiency virus. J. Virol. 2007, 81, 362–373.

Banerji, D. The placeof indigenous and Western systems of medicine in the health services of India. Soc. Sci. Med. [Med. Psychol. Med. Sociol.]. 1981, 15A(2), 109–114.

Barre-Sinoussi, F., Chermann, J. C., Rey, F., Nugeyre, M. T., Chamaret, S., Gruest, J., Dauguet, C., Axler-Blin, C., Vezinet-Brun, F., Rouzioux, C et al. Isolation of a T-lymphotropic retrovirus from a patient at risk for acquired immune deficiency syndrome (AIDS). Science. 1983, 220, 868–871.

Birk, T., McGrady, A., MacArthur, R., Khuder, S. The effects of massage therapy alone and in combination with other complementary therapies on immune system measures and quality of life in Human Immunodeficiency Virus. The Journal of Alternative and Complementary Medicine. 2000, 6(5), 405–414.

Blayne, C; Jessica, J. Vaginal microbicides and the prevention of HIV transmission. Lan. Inf. Dis. 2008, 8, 685–697.

Bryant, M. L., Ratner, L., Duronio, R. J., Kishore, N. S., Devadas, B., Adams, S. P., Gordon, J. I. Incorporation of 12-methoxydodecanoate into the human immunodeficiency virus 1 gag polyprotein precursor inhibits its proteolytic processing and virus production in a chronically infected human lymphoid cell line. Proc. Natl. Acad. Sci. USA. 1991, 88(6), 2055–2059.

Buchbinder, S. P., Mehrotra, D. V., Duerr, A et al. Efficacy assessment of a cell-mediated immunity HIV-vaccine trials (the STEP study): a double-blind, randomized, placebo-controlled, test-of-concept trial. Lancet. 2008, 372(9653), 1881–1893.

Budihas, S. R., Gorshkova, I., Gaidamakov, S., Wamiru, A., Bona, M. K., Parniak, M. A., Crouch, R. J., McMohan, J. B., Beutler, J. A., Le Grice, SFJ. Selective inhibition of HIV-1 reverse transcriptase-associated ribonuclease H activity by hydroxylated troplolones. Nuc. Acid Res. 2005, 33, 1249–1256.

Byrn, R. A., Mordenti, J., Lucas, C., Smith, D., Marsters, S. A., Johnson, J. S., Cossum, P., Chamow, S. M., Wurm, F. M., Gregory, T et al. Biological properties of a CD4 immunoadhesin. Nature. 1990, 344(6267):667–670.

Callahan, L. N., Phelan, M., Mallinson, M., Norcross, M. A. Dextran sulfate blocks antibody binding to the principal neutralizing domain of human immunodeficiency virus type 1 without interfering with gp120-CD4 interactions. J. Virol. 1991, 65(3), 1543–1550.

Carter, W. A., Strayer, D. R., Brodsky, I., Lewin, M., Pellegrino, M. G., Einck, L., Henriques, H. F., Simon, G. L., Parenti, D. M., Scheib, R. G. et al. Clinical, immunological, and virological effects of ampligen, a mismatched double-stranded RNA, in patients with AIDS or AIDS-related complex. Lancet. 1987, 1(8545), 1286–1292.

CDC. Kaposi's sarcoma and Pneumocystis pneumonia among homosexual men-New York City and California. MMWR Morb. Mortal Wkly. Rep. 1981, 30, 305–308.

Chatterjee P. HIV/AIDS prevention carries on in rural India. Lancet Infect. Dis. 2004, 4, 386.

Chatterjee P. Saving India's women from HIV/AIDS. Lancet Infect. Dis. 2004, 4, 714.

Chen, Z., Telfier, P., Gettie, A., Reed, P., Zhang, L., Ho, D. D, Marx, P. A. Genetic Characterization of new West African simian immunodeficiency virus SIVsm: Geographic clustering of household-derived SIV strains with human immunodeficiency virus type 2 subtypes and genetically diverse viruses from a single feral sooty mangabey troop. J. Virol. 1996, 70, 3617–3627.

Clavel, F., Guetard, D., Brun-Vezinet, F., Chamaret, S., Rey, M. A., Santos-Ferrerira, M. O., Laurent, A. G., Dauguet, C., Katlama, C., Rouzioux, C. et al. Isolation of a new human retrovirus from west African patients with AIDS. Science. 1986, 233, 343–346.

Coates, T. J., Collins, C. Preventing HIV infection. Sci. Am. 1998, 279, 96–97.

Coffin, J., Haase, A., Levy, J. A., Montagnier, L., Oroszlan, S., Teich, N., Temin, H., Toyoshima, K., Varmus, H., Vogt, P., Weiss, R. A. What to call the AIDS virus? Nature. 1986, 321(6065), 10.

Cohen, M. S., Shaw, G. M., McMichael, A. J., Haynes, B. F. Acute-HIV-1 Infection: Basic, clinical and public health perspectives. N. Eng. J. Med. 2011, 364, 1943–1954.

Conant, M., Goldstein, G., Hirsch, R. L., Meyerson, L. A., A. B. Kremer. Twenty-four week double blind evaluation of thymopentin treatment on disease progression in HIV infected patients, abstr. S.B.485. Abstr. 6th Int. Conf. AIDS, San Francisco. 1990.

Connor, E. M., Sperling, R. S., Gelber, R. et al. Reduction of maternal-infant transmission of human immunodeficiency virus type 1with zidovudine treatment. N. Eng. J. Med. 1994, 331, 1173–1180.

Correll, P. K., Law, M. G., McDonald, A. M., Cooper, D. A., Kaldor, J. M. HIV disease progression in Australia in the time of combination antiretroviral therapies. Med. J. Aust. 1998, 169, 469–472.

De Leys, R., Vanderborght, B., Vanden Haesevelde, M., Heyndrickx, L., van Geel, A., Wauters, C., Bernaerts, R., Saman, E., Nijs, P., Willems, B. et al. Isolation and partial characterization of an unusual human immunodeficiency retrovirus from two persons of west-central African origin. J. Virol. 1990, 64, 1207–1216.

de Silva, T. I., Cotton, M., Rowland-Jones, S. L. HIV-2: The forgotten AIDS virus. Trends Microbiol. 2008, 16, 588–595.

Deivanayagam, C. N., Krishnarajasekhar, O. R., Ravichandaran, N. Evaluation of Siddha medicare in HIV disease. J. Assoc. Physicians. India. 2001, 49, 390–391.

Diego, M., Field, T., Hernandez-Reif, M., Shaw, K., Friedman, L., Ironson, G.HIV adolescents show improved immune function following massage therapy. Intern. J. Neuroscience. 2001, 106, 35–45.

Dreyer, G. B., Metcalf, B. W., Tomaszek, T. A., Jr., Carr, T. J., Chandler III, A. C., Hyland, L., Fakhoury, S. A., Magaard, V. W., Moore, M. L., Strickler, J. E., Debouck, C., Meed, T. D. Inhibition of human immunodeficiency virus 1 protease *in vitro*: rational design of substrate analogue inhibitors. Proc. Natl. Acad. Sci. USA. 1989, 86, 9752–9756.

Elias, C., Coggins, C., Female-controlled methods to prevent sexual transmission of HIV. AIDS. 1996, 10, S43-S51.

Erickson, J., Neidhart, D. J., VanDrie, J., Kempf, D. J., Wang, X. C., Norbeck, D. W., Plattner, J. J., Rittenhouse, J. W., Turon, M., Wideburg, N., Kohlbrenner, W. E., Simmer, R., Helfrich, R., Paul, D. A., Mark, K. Design, activity, and 2.8 A crystal structure of a C2 symmetric inhibitor complexed to HIV-1 protease. Science. 1990, 249, 527–533.

Fauci, A. S. The AIDS epidemic. Considerations for the 21st Century. N. Engl. J. Med. 1999, 341, 1046–1050.

Gallo, R. C., Salahuddin, S. Z., Popvic, M., Shearer, G. M., Kaplan, M., Haynes, B. F., Palker, T. J., Redfield, R., Oleske, J., Safai, B. et al. Frequent detection and isolation of cytopathic retroviruses (HTLV-III) from patients with AIDS and at risk for AIDS. Science. 1984, 224, 500–503.

Gao, F., Bailes, E., Robertson, D. L. et al. Origin of HIV-1 in chimpanzee *Pan troglodytes troglodytes*. Nature. 1999, 397, 617–622.

Gao, F., Yu, L., White, A. T., Pappas, P. G., Barchure, J., Hanson, A. P., Greene, B. M., Sharp, P. M., Shaw, G. M., Hahn, B. H. Human infection by genetically diverse SIVsm-related HIV-2 in west Africa. Nature. 1992, 358, 495–499.

Gelman, B. B., Chen, T., Lisinicchia, J. G., Soukup, V. M., Carmical, J. R., Starkey, J. M., Masliah, E., Commins, D. L., Brandt, D., Grant, I., Singer, E. J., Levine, A. J., Miller, J., Winkler, J. M., Fox, H. S., Luxon, B. A., Morgello, S. The National NeuroAIDS Tissue Consortium brain gene array: two types of HIV-associated neurocognitive impairment. PLoS One. 2012, 7(9), e46178.

Glenn, G. M., Taylor, D. N., Li, X., Frankel, S., Montemarano, A., Alving, C. R. Transcutaneous immunization: a human vaccine delivery strategy using a patch. Nat. Med. 2000, 6, 1403–1406.

Go, V. F., Sethulakshmi, C. J., Bentley, M. E., Sivaram, S., Srikirishnan, A. K., Solomon, S., Celentano, D. D. When HIV-prevention messages and gender norms clash: the impact of domestic violence on women's HIV risk in slums of Chennai, India. AIDS Behav. 2003, 7, 263–272.

Government of India. Ministry of Health and Family Welfare: Financing and Delivery of Health care Services in India. 2005.Government of India: Indian Systems of medicine and Homeopathy. Annual report. [http://mohfw.nic.in/reports/Annual0506/Ayush%20 annual%20report%20final.pdf].

Gray, R. H., Kigozi, G., Serwadda, D et al. male Circumcision for HIV prevention in men in rakai, Uganda: a randomised trial. Lancet. 2007, 369, 657–666.

Greene, W. C. A history of AIDS: Looking back to see ahead. Eur. J. Immunol. 2007, 37 (Suppl 1), S94-S102.

Gurtler, L. G., Hauser, P. H., Eberle, J., von, B. A., Knapp, S., Zekeng, L., Tsague, J. M., Kaptue, L. A new subtype of human immunodeficiency virus type 1 (MVP-5180) from Cameroon. J. Virol. 1994, 68, 1581–1585.

Harbison, M. A., Gillis, J. M., Pinkston, P., Byrn, R. A., Rose, R. M., Hammer, S. M. Effects of recombinant soluble CD4 (rCD4) on HIV-1 infection of monocyte/macrophages. J. Infect. Dis. 1990, 161, 1–6.

Hersh, E. M., Brewton, G., Abrams, D., Bartlett, J., Galpin, J., Gill, P., Gorter, R., Gottlieb, M., Jonikas, J. J., Landesman, S., Levine, A., Marcel, A., Petersen, E. A., Whiteside, M., Zahradnik, J., Negron, C., Boutitie, F., Caraux, J., Dupuy, J. M., Salmi, L. R. Ditiocarb sodium (diethyldithiocarbamate) therapy in patients with symptomatic HIV infection in AIDS. JAMA. 1991, 265, 1538–1544.

Hirsch, V. M., Olmsted, R. A., Murphey-Corb, M., Purcell, R. H., Johnson, P. R. An African primate lentivirus (SIVsm) closely related to HIV-2. Nature. 1989, 339, 389–392.

Hladik, F., McElarth, M. J. Setting the stage: Host invasion by HIV. Nat. Rev. Immunol. 2008, 8, 447–457.

Ho, D. D. Rapid turnover of plasma virions and CD4 lymphocytes in HIV-1 infection. Nature. 1995, 373, 123–126.

Ho, D. D., Hartshorn, K. L., Rota, T. R., Andrews, C. A., Kaplan, J. C., Schooley, R. T., Hirsch, M. S. Recombinant human interferon alfa-A suppresses HTLV-III replication *in vitro*. Lancet i, 1985, 602–604.

Husain, M., D'Agati, V. D., He, J. C., Klotman, M. E., Klotman, P. E. HIV-1 Nef induces dedifferentiation of podocytes in vivo: a characteristic feature of HIVAN. AIDS. 2005, 19(17), 1975–80.

Husain, M., Singhal, P. C. HIV-1 entry into renal epithelia. J. Am. Soc. Nephrol. 2011, 22(3), 399–402.

Ishikawa, K., Janssens, W., Banor, J. S., Shinno, T., Piedade, J., Sata, T., Ampofo, W. K., Brandful, J. A., Koyanagi, Y., Yamamoto, N., et al. Genetic Analysis of HIV type 2 from Ghana and Guinea-Bissau, West Africa. AIDS Res. Hum. Retroviruses. 2001, 17, 1661–1663.

Jacks, T., Power, M. D., Masiarz, F. R., Luciw, P. A., Barr, P. J., Varmus, H. E. Characterization of ribosomal frameshifting in HIV-1 gag-pol expression. Nature (London). 1988, 331, 280–283.

James, J. K., Cecilia, A. M., Tracey, A. D., Peter, B. G., Steve, G. S., McElarth, M. J., Lawrence, C. HIV Vaccine trials Network: activities and achievements of the first decade and beyond. Clin.Invest. 2012, 2(3), 245–254.

Johnston, S. A., Tang, D. C. Gene gun transfection of animal cells and genetic immunization. Methods Cell Biol. 43 Part A, 1994, 353–365.

Jose, E. An HIV Vaccine: how and when? Bulleitin of World Health Organization. 2001, 79(12), 1133–1137.

Karpas, A., Fleet, G. W., Dwek, R. A., Petursson, S., Namgoong, S. K., Ramsden, N. G., Jacobs, G. S., Rademacher, T. W. Aminosugar derivatives as potential anti-human immunodeficiency virus agents. Proc. Natl. Acad. Sci. USA. 1989, 85, 9229–9233.

Kattumuri, R. One-and-a-half decades of HIV/AIDS in Tamil Nadu: how much do patients know now? Int. J. STD AIDS. 2003, 14, 552–559.

Kelly, C. G., Shattock, R. J. Specific microbicides in the prevention of HIV infection. Journal of Internal Medicine. 2011, 270, 509–519.

Klein, A. India: Supreme Court suspends manufacture of ayurvedic medicine being sold as a "cure" for AIDS. HIV AIDS Policy Law Rev. 2007, 12(1), 54.

Koen, K. A., Van, R., Purnima, M., Mirriam, R., Karl, K., Venkatesan, C., Durai, S. Empowering the people: Development of an HIV peer education model for low literacy rural communities in India. Human Resources for Health. 2008, 6, 1–11.

Lane, H. C., Davey, V., Kovacs, J. A., Feinberg, J., Metcalf, J. A., Herpin, B., Walker, R., Deyton, L., Jr., Davey, R. T., Falloon, J., Polis, M. A., Salzman, N. P., Baseler, M., Masur, H., Fauci, A. S. Interferon-alpha in patients with asymptomatic human immunodeficiency virus (HIV) infection. A randomized, placebo-controlled trial. Ann. Intern. Med. 1990, 112, 805–811.

Laurence, J., Sikder, S. K., Kulkosky, J., Miller, P., Tso, P. O. Induction of chronic human immunodeficiency virus infection is blocked *in vitro* by a methylphosphonate oligodeoxynucleoside targeted to a U3 enhancer element. J. Virol. 1991, 65, 213–219.

Levy, J. A. Pathogenesis of human immunodeficiency virus infection. Microbiol. Rev. 1993, 57, 183–289.

Man, L. Y., Yamina, B., Shu, Y. l. E., Kuan, T. J. siRNA, miRNA and HIV: Promises and challenges. Cell Research. 2005, 15(11–12), 935–946.

Mannar, R. M., Amit, K., Mohammad, A., Amir, A., Dioxovanadium (V) and μ-oxo bis[oxovanadium (V)] complexes containing thiosemicarbazone based ONS donor set and their antiamoebic activity. Inorganica Chimica Acta. 2006, 359(8), 2439–2447.

Mariano, E. Attenuated poxvirus vectors MVA and NYVAC as promising vaccine candidates against HIV/AIDS. Human vaccines. 2009, 5 (12), 867–871.

Matsukura, M., Shinozuka, K., Zon, G., Mitsuya, H., Reitz, M., Cohan, J. S., Broder, S. Phosphorothioate analogs of oligodeoxynucleotides: inhibitors of replication and cytopathic effects of human immunodeficiency virus. Proc. Natl. Acad. Sci. USA. 1987, 84, 7706–7710.

Matsukura, M., Zon, G., Shinozuka, K., Robert-Guroff, M., Shimada, T., Stein, C. A., Mitsuya, H., Wong-Staal, F., Cohen, J. S., Broder, S. Regulation of viral expression of human immunodeficiency virus *in vitro* by an antisense phosphorothioate oligodeoxynucleotide against rev (art/trs) in chronically infected cells. Proc. Natl. Acad. Sci. USA. 1989, 86, 4244–4248.

Mauclere, P., Loussert-Ajaka, I., Damond, F., Fagot, P., Souquieres, S., Monny, L. M., Mbopi, Keou, F. X., Barre-Sinoussi, F., Saragosti, S., Brun-Vezint, F., et al. Serological and Virological Characterization of HIV-1 group O infection in Cameroon. AIDS. 1997, 11, 445–453.

Maulik, P. K., Darmstadt, G. L. Community-based interventions to optimize early childhood development in low resource settings. Journal of Perinatology. 2009, 29, 531–542.

McEnery, R. Vaccine candidate targeting dendritic cells enters clinical trial. IAVI Rep. 2010, 14(4), 23.

McGovern, S. L., Caselli, E., Grigorieff, N., Shoichet, B. K. A common mechanism underlying promiscuous inhibitors from virtual and high-throughput screening. J. Med. Chem. 2002, 45(8), 1712–1722.

McGrath, M. S., Hwang, K. M., Caldwell, S. E., Gaston, I., Luk, K. C., Wu, P., Ng, V. L., Crowe, S., Daniels, J., Marsh, J., Deinhart, T., Lekas, P. V., Vennari, J. C., Yeung, H. W., Lifson, J. D. GLQ223: an inhibitor of human immunodeficiency virus replication in acutely and chronically infected cells of the lymphocyte and mononuclear phagocyte lineage. Proc. Natl. Acad. Sci. USA. 1989, 86, 2844–2848.

McQuade, T. J., Tomasselli, A. G., Liu, L., Karacostas, V., Moss, B., Sawyer, T. K., Heinrikson, R. L., Tarpley, W. G. A synthetic HIV-2 protease inhibitor with antiviral activity arrests HIV-like particle maturation. Science. 1990, 247, 454–456.

Meek, T. D., Lambert, D. M., Dreyer, G. B., Carr, T. J., Tomaszek, T. A., Jr., Moore, M. L., Strickler, J. E., Debouck, C., Hyland, L. J., Matthews, T. J., Metcalf, B. W., Petteway, S. R. Inhibition of HIV-1 protease in infected T-lymphocytes by synthetic peptide analogues. Nature (London). 1990, 343, 90–92.

Meruelo, D., Lavaie, G., Lavie, D. Therapeutic agents with dramatic antiretroviral activity and little toxicity at effective doses: aromatic polycyclic diones hypericin and pseudohypericin. Proc. Natl. Acad. Sci. USA. 1988, 85, 5230–5234.

Michaels, S. H., Clark, R., Kissinger, P. Declining morbidity and mortality among patients with advanced human immunodeficiency virus infection. N. Engl. J. Med. 1998, 339, 405–406.

Mofenson, L. M. Short-course zidovudine for prevention of perinatal infection. Lancet. 1999, 353, 766–767.

Montefiori, D. C., Robinson, W. E. Jr.; Mitchell, W. M. *In vitro* evaluation of mismatched double-stranded RNA (ampligen) for combination therapy in the treatment of acquired immunodeficiency syndrome. AIDS Res. Hum. Retroviruses. 1989, 5, 193–203.

Moore, J. P., McKeating, J. A., Weiss, R. A., Sattentau, Q. J. Dissociation of gpl20 from HIV-1 virions induced by soluble CD4. Science. 1990, 250, 1139–1142.

Palca, J. Promising AIDS drug looking for a sponsor. Science. 1991, 253, 262–263.

Paul, M. S., Beatrice, H. H. Origins of HIV and the AIDS Pandemic. Cold Spring Harbor Perspectives in Medicine. 2011, 1, a006841.

Peeters, M., Gueye, A., Mboup, S., Bipollet-Ruche, F., Ekaza, E., Mulanga, C., Ouedrago, R., Gandji, R., Mpele, P., Dibanga, G et al. Geographical distribution of HIV-1 group O viruses in Africa. AIDS. 1997, 11, 493–498.

Pejchal, R., Wilson, I. A. Structure-based vaccine design in HIV: blind men and the elephant? Curr. Pharm. Des. 2010, 16(33), 3744–3753.

Pieniazek, D., Ellenberger, D., Janini, L. M., Ramos, A. C., Nkengasong, J., Sassan-Morokro, M., Hu, D. J., Coulibally, I. M., Ekpini, E., Bandea, C et al. Predominance of human immunodeficiency virus type 2 subtype B in Abidjan, Ivory Coast. AIDS Res. Hum. Retroviruses. 1999, 15, 603–608.

Pilcher, C. D., Tien, H. C., Eron, J. J. et al. Brief but efficient: acute infection and the sexual transmission of HIV. J. Infect. Dis. 2004, 189, 1785–1792.

Plantier, J. C., Leoz, M., Dickerson, J. E., De Oliveira, F., Cordonnier, F., Lemee, V., Damond, F., Robertson, D. L., Simon, F. A new human immunodeficiency virus derived from gorillas. Nature Med. 2009, 15, 871–872.

Popovic, M., Sarngadharan, M. G., Read, E., Gallo, R. C. Detection, isolation, and continuous production of cytopathic retroviruses (HTLV-III) from patients with AIDS and pre-AIDS. Science. 1984, 224, 497–500.

Powers, K. A., Poole, C., Pettifor, A. E., Cohen, M. S. Rethinking the heterosexual infectivity of HIV-1: a systematic review and meta-analysis. Lan. Inf. Dis. 2008, 8, 553–563.

Ramachandran, S et al. Knowledge, Attitudes, and Practices of Antiretroviral therapy Among Adults Attending Private and Public Clinics in India. AIDS Patient Care STDS. 2007, 21(2), 129–142.

Ratner, L., vander, H. N., Dedera, D. Inhibition of HIV and SIV infectivity by blockade of alpha-glucosidase activity. Virology. 1991, 181, 180–192.

Reisinger, E. C., Kern, P., Ernst, M., Bock, P., Flad, H. D., Dietrich, M., and the German DTC Study Group. Inhibition of HIV progression by dithiocarb. Lancet. 1990, 335, 679–82.

Roberts, N. A., Martin, J. A., Kinchington, D., Broadhurst, A. V., Craig, J. C., Duncan, I. B., Galpin, S. A., Handa, B. K., Kay, J., Krohn, A., Lambert, R. W., Merrett, J. H., Mills, J. S., Parkes, K. E. B., Redshaw, S., Ritchie, A. J., Taylor, D. L., Thomas, G. J., Machin, P. J. Rational design of peptidebased HIV proteinase inhibitors. Science. 1990, 248, 358–361.

Robertson, D. L., Anderson, J. P. ; Bradac, J. A., et al. HIV-1 nomenclature proposal. Science. 2000, 288(5463), 55–56.

Rowland-Jones, S. L., Whittle, H. C. Out of Africa: What can we learn from HIV-2 about protective immunity to HIV-1? Nat. Immunol. 2007, 8, 329–331.

Saba, J. The results of the PETRA intervention trial to prevent intervention trial to prevent perinatal transmission in sub-Sahara Africa. Chicago: Foundation for Retrovirology and Human Health, 1999 (See http://www.retroconference.otg/99/lect_symposia/sym_session8.htm).

Santiago, M. L., Range, F., Keele B, F., Li, Y., Bailes E, Bibollet-Ruche, F., Fruteau, C., Noe, R., Peeters, M., Brookfield, J. F. et al. Simian immunodeficiency virus infection in free-ranging sooty mangabeys (*Cercocebus atys atys*) from the Tai forest, Cote d'Ivoire: Implications for the origin of epidemic human immunodeficiency virus type 2. J. Virol. 2005, 79, 12515–12527.

Sarver, N., Cantin, E. M., Chang, P. S., Zaia, J. A., Lande, P. A., Stephens, D. A., Rossi, J. J. Ribozymes as potential anti-HIV-1 therapeutic agents. Science. 1990, 247, 1222–1225.

Scott, G. K., Jerome, A. Z. Stem Cell-Based Approaches to treating HIV Infection. Curr opin HIV AIDS. 2011, 6, 68–73.

Sendi, P. P., Bucher, H. C., Craig, B. A., Pfluger, D., Battegay, M. Estimating AIDS free survival in a severely immunosuppressed asymptomatic HIV- infected population in the era of antiretroviral triple combination therapy. Swiss HIV Cohort study. J. Acquired Immune Defic. Syndr. Hum. Retrovirol. 1999, 20, 376–381.

Shor-Posner, G., Hernandez-reif, M., Miguez, M., Fletcher, M., Quintero, N., Baez, J., Perez-then, E., Soto, S., Mendoza, R., Castillo, R., Zhang, G. Impact of a Massage therapy clinical trial on immune status in young Dominician children infected with HIV-1. The Journal of Alternative and Complementary Medicine. 2006, 12(6), 511–516.

Simon, F., Mauclere, P., Roques, P., Lousset-Ajaka, I., Muller-Trutwin, M. C., Saragosti, S., Georges-Coubort, M. C., Barre-Sinoussi, F., Brun-Vezinet, F. Identification of a new human immunodeficiency virus type 1 distinct from group M and group, O. Nat. Med. 1998, 4, 1032–1037.

Singh, P., Yadav, R., Pandey, A. Utilization of indigenous systems of medicine and homeopathy in India. Indian, J. Med. Res. 2005, 122(2), 137–142.

Sivaram, S., Johnson, S., Bentley, M. E., Go, V. F., Latkin, C., Srikrishan, A. K., Celentano, D. D., Solomon, S. Sexual health promotion in Chennai, India: key role of communication among social networks. Health Promot. Int. 2005, 20, 327–333.

Smith, D. H., Byr, R. A., Marsters, S. A., Gregory, T., Groopman, J. E., Capon, D. J. Blocking of HIV-1 infectivity by a soluble, secreted form of the CD4 antigen. Science. 1987, 238, 1704–1707.

Staib, C., Kisling, S., Erfle, V., Sutter, G., Inactivation of the viral interleukin 1 beta receptor improves CD8+ T-cell memory responses elicited upon immunization with modified vaccinia virus Ankara. J. Gen. Virol. 2005, 86, 1997–2006.

Stein, C. A., Matsukura, M., Subasinghe, C., Broder, S., Cohen, J. S. Phosphorothioate oligo deoxynucleotides are potent sequence nonspecific inhibitors of de novo infection by HIV. AIDS Res. Hum. Retroviruses. 1989, 5, 639–646.

Susan, L. H., Quinette, L., Linzette, M., Jeanine, U., Sue, S. Massage therapy for people with HIV/AIDS. Cochrane Database of Systematic Reviews. 2010, 1, 1–23.

Takahashi, I., Nakanishi, S., Kobayashi, E., Nakano, H., Suzuki, K., Tamaoki, T. Hypericin and pseudohypericin specifically inhibit protein kinase C: possible relation to their antiretroviral activity. Biochem. Biophys. Res. Commun. 1989, 165, 1207–1212.

Tarapdhar, P., Rray, T. G., Haldar, D., Chatterjee, A., Dasgupta, A., Saha, B., Malik, S. Socioeconomic consequences of HIV/AIDS in the family system. Niger Med. J. 2011, 52(4), 250–253.

Tartaglia, J., Perkus, M. E., Taylor, J., Norton, E. K., Audonnet, J. C., Cox, W. I. et al., NYVAC: a highly attenuated strain of vaccinia virus. Virology. 1992, 188, 217–232.

Thompson, S. E., Calabrese, L., Hirsch, R. L., Longworth, D., Rehm, S., Kremer, A. B., Meyerson, L. A., Goldstein, G. Effects of thymopentin on disease progression and surrogate markers in HIV infection-a 1 year study, abstr. S.B.484. Abstr. 6th Int. Conf. AIDS, San Francisco. 1990.

Vallari, A., Bodelle, P., Ngansop, C., Makamche, F., Ndembi, N., Mbanya, D., Kaptue, L., Gurtler, L. G., McArthur, C. P., Devare, S. G. et al. Four new HIV-1 group N isolates from Cameroon: Prevalence continues to be low. AIDS Res. Hum. Retroviruses. 2010, 26, 109–115.

Vallari, A., Holzmayer, V., Harris, B., Yamaguchi, J., Ngansop, C., Makamche, F., Mbanya, D., Kaptue, L., Ndembi, N., Gurtler, L. et al. Confirmation of putative HIV-1 group P in Cameroon. J. Virol. 2011, 85, 1403–1407.

Van, R. K. K. A. Motor-biking through rural India on an HIV mission. AIDS. 2004, 18, N13–N18.

Vanessa, P., Nina, T., Jeffrey, M., Jacobson, B. W., Fred, C. K. Combinatorial Approaches to the Prevention and Treatment of HIV-1 infection. Antimicrobial Agents and Chemotherapy. 2011, 55(5), 1831–1842.

Vincent, P., Andrew, B. Essential Roles for Dendritic Cells in the pathogenesis and potential treatment of HIV disease. The, J. of Invest. Derm. 2002, 365–369.

Wei, X. Viral Dynamics in Human immunodeficiency virus type 1 infection. Nature. 1995, 373, 117–122.

Weiss, R. A., Wrangham, R. W. From Pan to epidemic. Nature. 1999, 397, 385–386.

Williams, A., Selwyn, P., Liberti, L., Molde, S., Njike, V., McCorkle, R., Zelterman, D., Katz, D. A randomized controlled trial of meditation and massage effects on quality of life in people with late-stage disease: a pilot study. Journal of Palliative Medicine. 2005, 8(5), 939–953.

World Health Organization: Traditional Medicine Strategy: 2002–2005. 2002 [http://whqlibdoc.who.int/hq/2002/WHO EDM TRM 2002.I.pdf].

World Health Organization: Traditional Medicine: Report by the Secretariat. Geneva: World Health Organization; 2003.

Yi, H. A., Fochtman, B. C., Rizzo, R. C., Jacobs, A. Inhibition of HIV entry by targeting the Envelope transsmembrane subunit gp41. Curr. HIV Res. 2016, 14, 283–294.

Youcef, M., Erik, D. C. Twenty-Six Years of Anti-HIV Drug Discovery: Where Do We Stand and Where Do We Go? Journal of Medicinal Chemistry. 2009, A-R.

Yu, F., Liu, X., Zhan, P., De, C. E. Recent advances in the research of HIV-1 RNase H inhibitors. Mini-Rev. Med. Chem. 2008, 8, 1243–1251.

CHAPTER 6

ANALYSIS OF A SIMPLE HIV/TB COINFECTION MODEL WITH THE EFFECT OF SCREENING

MINI GHOSH

CONTENTS

ABSTRACT

This work presents a non-linear mathematical model to study the transmission dynamics of 'Human Immunodeficiency Virus (HIV)' and 'Tubercle Bacillus (TB)' co-infection incorporating the effect of screening of both the HIV and TB infected individuals. The basic reproduction numbers corresponding to both the HIV and TB are obtained and it is shown that the disease free equilibrium is stable only when both the reproduction numbers are less than one. The existence and stability of TB-only and HIV-only equilibria are discussed in-detail. In this work, we observe that the co-infection equilibrium point exists only under some restrictions on the parameters provided both the basic reproduction numbers are greater than one. Importantly, this equilibrium is always unstable. A numerical simulation is reported to support and strengthen the major analytical findings. Furthermore, the effects of screening of the infectives on the equilibrium level of HIV and TB infected population are investigated and it is found that the screening with proper counseling results in a significant reduction in the number of HIV and TB infected individuals.

6.1 INTRODUCTION

The 'Acquired Immune Deficiency Syndrome (AIDS)' is a deadly disease of the human immune system that is caused by infection with the 'Human Immunodeficiency Virus (HIV).' As per the current understanding of the AIDS, in the initial stage of infection, a person experiences a brief period of influenza-like illness. Later in the subsequent stages, this is typically followed by a prolonged period without any significant symptoms. As the illness progresses, it interferes more and more with the immune system, making the person much more likely to get infections, including opportunistic infections and tumors that do not usually affect people who have efficient working immune systems. Worldwide about 33.2 million people live with AIDS and about 2.1 million AIDS related death occurs including 3,30,000 children (UNAIDS, 2007).

Though a breakthrough has been reported recently (for the details see (World Health Organization HIV/TB Facts, 2011), so far there is no cure for AIDS and this disease is endemic in many part of the World and especially in sub-Saharan Africa. The old wisdom of 'prevention is better than cure' is truly applicable in the case of HIV/AIDS transmission.

HIV transmission can be reduced if preventive measures are taken well in advance. This is an important factor in screening of infectives and to provide proper counseling so that further transmissions due to that particular infective can be prevented. And it is observed that HIV infected individuals are more likely to get other infections and diseases, as well. Out of these diseases, the 'Tubercle Bacillus (TB)' is a common, and in many cases lethal, infectious disease that is caused by various strains of 'mycobacteria' usually Mycobacterium tuberculosis. In general, the TB typically attacks the lungs, but can also affect other parts of the body. As per current understanding, it is spread through the air when people who have an active TB infection cough, sneeze, or otherwise transmit respiratory fluids through the air. Normally, most of the infections are asymptomatic and latent, but about one in ten latent infections eventually progresses to active disease which, if left untreated, kills more than 50% of those so infected. The AIDS and TB exhibit some sort of synergistic relationship, where each accelerates the progression of the other. TB is the main cause of death among people living with HIV (Pollack and Mcneil, 2013). So it is highly desirable to incorporate the effect of TB infection while studying the dynamics of HIV/AIDS to minimize the casualties caused by the HIV/AIDS epidemic.

In our opinion, there is an urgent need to frame the effective strategies that would reduce the transmission rates of these diseases. Mathematical modeling plays a very important role in understanding and control of infectious diseases. There are several mathematical models that describes the transmission dynamics of HIV/AIDS and HIV-TB co-infections, (May and Anderson, 1987, 1988; Rao, 2003; Sharomi et al., 2008; Lih-ing et al., 2009; Naresh et al., 2009; Yan, 2010; Yang et al., 2011; Naresh et al., 2011; Ermentrout, 2002). Most of these models do not incorporate the effects of screening of infectives especially the HIV-TB coinfection models. In our work we consider a simple HIV-TB co-infection model by incorporating the screening of infectives. Here screening of both HIV and TB infectives are taken into account and it is assumed that screening is associated with proper counseling. Here we assume that the screened HIV infectives are not taking part in the transmission of HIV. However, some screened HIV infectives may contribute in the transmission of HIV virus, but in this work we have ignored those individuals as if they are not screened. Additionally, it is assumed that counseling is provided to screened/detected TB patients so that they are aware of consequences of HIV-TB co-infection, so they keep themselves away from HIV infectives.

6.2 THE MODEL

Here the population under consideration is the adult population as nearly 80% HIV transmission is due to hetero-sexual transmission. The total population *N(t)* at time *t* is subdivide into mutually-exclusive compartments, namely susceptible (*S(t)*), individuals infected with TB ($I_1(t)$) assumed to be infectious, HIV/AIDS infected individuals ($I_2(t)$), dually infected individuals i.e. infected with active TB and HIV/AIDS ($I_3(t)$). It is assumed that there are several screening centers where screening of both HIV and TB are performed. Health providers also provide proper counseling to individuals who are identified as HIV or TB infected. So it is assumed that a fraction of HIV infected individuals is not taking part in the transmission of HIV. Additionally it is assumed that detected TB patients are very careful and are keeping themselves away from HIV infection. It is also assumed that the co-infected individuals become very sick and they do not contribute in the transmission of HIV or TB. Keeping in view all the above facts a mathematical model can be formulated as follows:

$$\frac{dS}{dt} = \Lambda - \beta_1 S I_1 - \beta_2 (1-\alpha) S I_2 - \mu S + \nu\eta I_1$$

$$\frac{dI_1}{dt} = \beta_1 S I_1 - \beta_2 (1-\alpha)(1-\eta) I_1 I_2 - (\nu\eta + \mu + \mu_1) I_1$$

$$\frac{dI_2}{dt} = \beta_2 (1-\alpha) S I_2 - \beta_1 I_1 I_2 - (\mu + \mu_2) I_2 \tag{1}$$

$$\frac{dI_3}{dt} = \{\beta_1 + \beta_2 (1-\alpha)(1-\eta)\} I_1 I_2 - (\mu + \mu_3) I_3$$

$$S(0) > 0, I_1(0) \geq 0, I_2(0) \geq, 0, I_3(0) \geq 0, I_4(0) \geq 0.$$

The parameters used in this model are as follows: Λ is the recruitment rate constant; α is the fraction of total HIV/AIDS infected individuals who are screened and are under proper counseling; η is the fraction of total TB infected individuals who are screened; ν is recovery rate of detected TB patients; μ is the natural death rate constant; β_1 is the rate of TB transmission, β_2 is the rate of HIV transmission; μ_1 is TB related death rate; μ_2 is HIV/AIDS related death rate; μ_3 is the death rate constant of co-infected individuals. Here all the parameters of the model are non-negative and it is noted that the rate of transmission of tuberculosis, i.e., β_1 is greater than the rate of transmission of HIV i.e., β_2 The region of attraction of this model is given by $D = \{(S, I_1, I_2, I_3) \in \mathbb{R}^4_+ : N = S + I_1 + I_2 + I_3 \leqslant \frac{\Lambda}{\mu}\}$ and the solutions $S(t), I_1(t), I_2(t)$ and $I_3(t)$ of the system (2.1) remains positively invariant. The transfer diagram of the model is described in Figure 6.1. The Basic reproduction number corresponding to TB and HIV are computed as $R_1 = \frac{\Lambda\beta_1}{\mu(\nu\eta + \mu + \mu_1)}$ and $R_2 = \frac{\Lambda\beta_2(1-\alpha)}{\mu(\mu+\mu_2)}$, respectively. The basic reproduction number is nothing but the number of new cases (infectives) generated from a single infective in his/her whole infectious period. And in general if this number is less than one then disease automatically dies out.

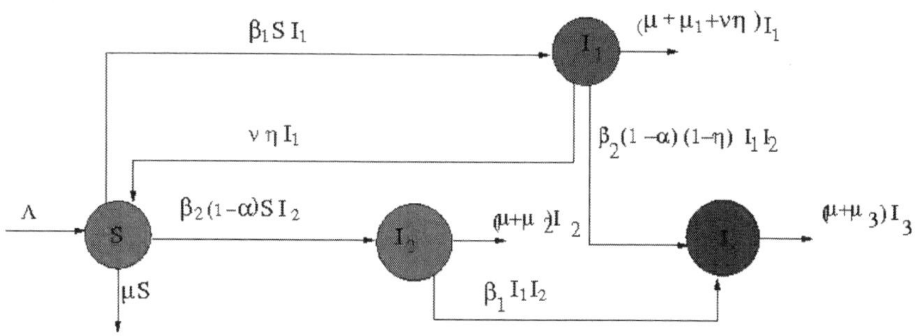

FIGURE 6.1 Transfer diagram of the model (Eq. (1)).

6.3 EQUILIBRIUM ANALYSIS

The equation system (Eq. (1)) has following disease free and boundary equilibria:

1. Disease free equilibrium $E_0 = (\frac{\Lambda}{\mu}, 0, 0, 0)$.

2. TB-only equilibrium where $E_1 = (S^*, I_1^*, 0, 0)$

$$S^* = \frac{\nu\eta + \mu + \mu_1}{\beta_1}, I_1^* = \frac{(\nu\eta + \mu + \mu_1)\mu}{\beta_1(\mu + \mu_1)}(R_1 - 1).$$

3. HIV-only equilibrium where $E_2 = (S^{**}, 0, I_2^{**}, 0)$

$$S^{**} = \frac{\mu + \mu_2}{\beta_2(1-\alpha)}, I_2^{**} = \frac{\mu(R_2 - 1)}{\beta_2(1-\alpha)}.$$

To find the coinfection equilibrium point, we equate the right hand side of the system (Eq. (1)) to zero which gives $I_3 = \frac{\{\beta_1 + \beta_2(1-\alpha)(1-\eta)\}I_1 I_2}{(\mu + \mu_3)}$, $I_1 = \frac{\beta_2(1-\alpha)S - (\mu + \mu_2)}{\beta_1}, I_2 = \frac{\beta_1 S - (\eta\nu + \mu + \mu_1)}{\beta_2(1-\alpha)(1-\eta)}$ and S is the root of the following quadratic equation

$$AS^2 + BS + C = 0 \tag{2}$$

Here $A = \beta_2(1-\alpha) + \frac{\beta_1}{1-\eta} > 0, B = -\left(\mu_2 + \frac{\nu\eta(1-\alpha)\beta_2}{\beta_1} + \frac{\nu\eta + \mu + \mu_1}{1-\eta}\right) < 0,$

$C = \frac{\nu\eta(\mu + \mu_2)}{\beta_1} - \Lambda$, and

$$B^2 - 4AC = \left[\left\{\mu_2 + \frac{\mu}{1-\eta}\right\} - \left\{\frac{\nu\eta + \mu_1}{1-\eta} + \frac{\nu\eta(1-\alpha)\beta_2}{\beta_1}\right\}\right]^2 + 4\Lambda\left\{(1-\alpha)\beta_2 + \frac{\beta_1}{1-\eta)}\right\}$$
$$+ \frac{4\mu_1\mu_2}{1-\eta} + \frac{4\mu\mu_1}{(1-\eta)^2} + \frac{4\mu\nu\eta^2}{(1-\eta)^2} + \frac{4\mu\nu\eta\beta_2(1-\alpha)}{\beta_1}\left(\frac{\eta}{1-\eta}\right) > 0.$$

Now if $C > 0$, then we get two positive roots of the quadratic equation (Eq. (2)), but we need to verify that whether both the roots satisfy the conditions corresponding to I_1 and I_2 to be positive or not? For I_1 and I_2 to be positive we want the following:

$$S > \max\left\{\frac{\mu+\mu_2}{\beta_2(1-\alpha)}, \frac{\nu\eta+\mu+\mu_1}{\beta_1}\right\}. \tag{3}$$

Additionally, our smaller root of the quadratic equation will be less than $\frac{-B}{2A}$, which gives

$$S < \frac{\left[\mu_2 + \frac{\mu}{1-\eta} + \left\{\frac{\nu\eta+\mu_1}{1-\eta} + \frac{\nu\eta(1-\alpha)\beta_2}{\beta_1}\right\}\right]}{2\left[(1-\alpha)\beta_2 + \frac{\beta_1}{1-\eta}\right]}$$

$$or\ S < \frac{\mu_2 + \frac{\mu}{1-\eta}}{2\left\{(1-\alpha)\beta_2 + \frac{\beta_1}{1-\eta}\right\}} + \frac{\left\{\frac{\nu\eta+\mu_1}{1-\eta} + \frac{\nu\eta(1-\alpha)\beta_2}{\beta_1}\right\}}{2\left\{(1-\alpha)\beta_2 + \frac{\beta_1}{1-\eta}\right\}} \tag{4}$$

Now we have following two cases: (i) when $\frac{\mu_2 + \frac{\mu}{1-\eta}}{\left\{(1-\alpha)\beta_2 + \frac{\beta_1}{1-\eta}\right\}} > \frac{\left\{\frac{\nu\eta+\mu_1}{1-\eta} + \frac{\nu\eta(1-\alpha)\beta_2}{\beta_1}\right\}}{\left\{(1-\alpha)\beta_2 + \frac{\beta_1}{1-\eta}\right\}}$, we get $S < \frac{\mu_2 + \frac{\mu}{1-\eta}}{\left\{(1-\alpha)\beta_2 + \frac{\beta_1}{1-\eta}\right\}}$. Also from Eq. (3), we get $S > \frac{\mu+\mu_2}{\beta_2(1-\alpha)}$. Thus we must have the following:

$$\frac{\mu+\mu_2}{\beta_2(1-\alpha)} < \frac{\mu_2+\dfrac{\mu}{1-\eta}}{\left\{(1-\alpha)\beta_2+\dfrac{\beta_1}{1-\eta}\right\}},$$

$$\Rightarrow \mu_2(1-\alpha)\beta_2+\frac{\mu\beta_2\alpha}{1-\eta}+\frac{\mu_2\beta_2}{1-\eta}<0,$$

which is not possible as all parameters are positive and α and η are less than one. Hence in this case we do not have positive I_1 corresponding to the smaller root (S) of the quadratic equation.

(ii) when $\dfrac{\mu_2+\dfrac{\mu}{1-\eta}}{\left\{(1-\alpha)\beta_2+\dfrac{\beta_1}{1-\eta}\right\}} < \dfrac{\left\{\dfrac{v\eta+\mu_1}{1-\eta}+\dfrac{v\eta(1-\alpha)\beta_2}{\beta_1}\right\}}{\left\{(1-\alpha)\beta_2+\dfrac{\beta_1}{1-\eta}\right\}}$,

we get $S<\dfrac{\left\{\dfrac{v\eta+\mu_1}{1-\eta}+\dfrac{v\eta(1-\alpha)\beta_2}{\beta_1}\right\}}{\left\{(1-\alpha)\beta_2+\dfrac{\beta_1}{1-\eta}\right\}}$. From the inequality Eq. (3), we want $S>\dfrac{v\eta+\mu+\mu_1}{\beta_1}$ Hence we must have

$$\frac{v\eta+\mu+\mu_1}{\beta_1} < \frac{\left\{\dfrac{v\eta+\mu_1}{1-\eta}+\dfrac{v\eta(1-\alpha)\beta_2}{\beta_1}\right\}}{\left\{(1-\alpha)\beta_2+\dfrac{\beta_1}{1-\eta}\right\}}.$$

This gives $\dfrac{(1-\alpha)\beta_2(\mu+\mu_1)}{\beta_2}+\dfrac{\mu}{1-\eta}<0$, which is false as all the parameters are positive and $\alpha<1$Thus, in this case I_2 becomes negative.

Hence we conclude that smaller root of the quadratic equation never gives both I_1 and I_2 positive. So for $C > 0$,we get unique interior (co-infection) equilibrium corresponding to the larger root of the quadratic equation provided the condition mentioned in the inequality equation (Eq. (3)) is satisfied.

And if C is negative, i.e. $\frac{\Lambda\beta_1}{\nu\eta(\mu+\mu_2)} > 1$, then we get unique positive root and corresponding to this we get unique interior (coinfection) equilibrium point provided condition mentioned in Eq. (3) is satisfied. Additionally, the S corresponding to the interior (co-infection) equilibrium point must be less than the S corresponding to the disease free equilibrium point, so we must have $S < \frac{\Lambda}{\mu}$. And using the inequality (3) we get $R_1 > 1$ and $R_2 > 1$. Thus we have unique endmeic cquilibrium point under the condition stated in Eq. (3) irrespective of $C < 0$ or $C > 0$ whenever both $R_1 > 1$ and $R_2 > 1$. Let us name this equilibrium as $E_3(\hat{S}, \hat{I}_1, \hat{I}_2, \hat{I}_3)$.

6.4 STABILITY ANALYSIS

The local asymptotic stability of the equilibria E_0, E_1, E_2, and E_3 are established using variational matrix method and stated in the following theorems.

Theorem 4.1. If $R_1 < 1$ and $R_2 < 1$, the disease free equilibrium E_0 is locally asymptotically stable and is unstable for $R_1 > 1$ or $R_2 > 1$.

Proof: To study the stability of disease free equilibrium the variational matrix M_0 of the system corresponding to disease free equilibrium E_0 is obtained as

$$M_0 = \begin{pmatrix} -\mu & -\beta_1\frac{\Lambda}{\mu}+\nu\eta & -\beta_2(1-\alpha)\frac{\Lambda}{\mu} & 0 \\ 0 & \beta_1\frac{\Lambda}{\mu}-(\mu+\mu_1+\eta\nu) & 0 & 0 \\ 0 & 0 & \beta_2(1-\alpha)\frac{\Lambda}{\mu}-(\mu+\mu_2) & 0 \\ 0 & 0 & 0 & -(\mu+\mu_3) \end{pmatrix}$$

The eigenvalues of this variational matrix are $\mu, \beta_1 \frac{\Lambda}{\mu} - (\mu + \mu_1 + \eta\nu$. Clearly here two eigenvalues are positive for $R_1 > 1$ and $R_2 > 1$ which implies instability of disease free equilibrium E_0. So the equilibrium is locally asymptotically stable provided $R_1 < 1$ and $R_2 < 1$.

Theorem 4.2. The TB only equilibrium E_1 is locally asymptotically stable provided $R_1 < R^*$ and unstable otherwise. The expression for R^* is given in the proof of the theorem.

Proof: To study the stability of TB only equilibrium the variational matrix M_1 of the system at E_1 is obtained as follows:

$$M_1 = \begin{pmatrix} -\beta_1 I_1^* - \mu & -\beta_1 S^* + \nu\eta & -\beta_2(1-\alpha)S^* & 0 \\ \beta_1 I_1^* & 0 & -\beta_2(1-\alpha)(1-\eta)I_1^* & 0 \\ 0 & 0 & \beta_2(1-\alpha)S^* - (\mu+\mu_2) - \beta_1 I_1^* & 0 \\ 0 & 0 & \{\beta_1 + \beta_2(1-\alpha)(1-\eta)\}I_1^* & -(\mu_3+\mu) \end{pmatrix}$$

where $S^* = \dfrac{\gamma+\mu+\mu_1}{\beta_1}, I_1^* = \dfrac{(\gamma+\mu+\mu_1)\mu}{(\mu+\mu_1)\beta_1}(R_1 - 1)$.

Clearly, two eigenvalues of this matrix M_1 are $-(\mu + \mu_3)$ and $\beta_2(1-\alpha)S^* - \beta_1 I_1^* - (\mu_1 + \mu_2)$. Other two eigenvalues are given by the roots of the following quadratic equation: $\psi^2 + (\beta_1 I_1^* + \mu)\,\psi + \beta_1 I^*(\mu + \mu_1) = 0$.

The above quadratic will have negative roots or will have roots with negative real parts as all the coefficients of the quadratic are positive. Hence, three eigenvalues of the matrix M_1 are either negative or have negative real parts. The fourth eigenvalue will be negative provided $\beta_2(1-\alpha)S^* - \beta_1 I_1^* - (\mu_1 + \mu_2) < 0$, which corresponds to $R_1 < R^*$, where

$$R^* = \left(\frac{\mu+\mu_1}{\mu}\right)\left\{\frac{\beta_2(1-\alpha)}{\beta_1} - \frac{(\mu+\mu_2)}{(\eta\nu+\mu+\mu_1)}\right\} + 1 = R^*.\,(say)$$

Hence the equilibrium point E_1 is locally asymptotically stable provided the condition stated in the theorem is satisfied.

Theorem 4.3. The HIV only equilibrium E_2 is locally asymptotically stable provided $R_1 > R^{**}$ otherwise it is unstable. The expression for R^{**} is given in the proof of this theorem.

Proof: To study the stability of HIV only equilibrium the variational matrix M_2 of the system corresponding to HIV only equilibrium E_2

$$M_2 = \begin{pmatrix} -\beta_2(1-\alpha)I_2^{**}-\mu & -\beta_1 S^{**}+\nu\eta & -\beta_2(1-\alpha)S^{**} & 0 \\ 0 & m_{22} & 0 & 0 \\ \beta_2(1-\alpha)I_2^{**} & -\beta_1 I_2^{**} & 0 & 0 \\ 0 & \{\beta_1+\beta_2(1-\alpha)(1-\eta)\}I_2^{**} & 0 & -(\mu_3+\mu) \end{pmatrix},$$

where $m_{22} = \beta_1 S^{**} - \beta_2(1-\alpha)(1-\eta)I_2^{**} - (\mu + \mu_1 + \eta\nu)$. Two eigenvalues of this variational matrix are $-(\mu + \mu_3)$, $\beta_1 S^{**} - \beta_2(1-\alpha)(1-\eta)I_2^{**} - (\eta\nu + \mu + \mu_1)$ and other two eigenvalues are roots of quadratic which are given by

$$\psi^2 + \{\beta_2(1-\alpha)I_2^{**} + \mu\}\psi + \beta_2^2(1-\alpha)^2 S^{**} I_2^{**} = 0 .$$

Clearly, the roots of this quadratic equation are either negative or have negative real parts as all the coefficients of the quadratic are positive. The equilibrium E_2 will be locally asymptotically stable provided all the eigenvalues are negative or have negative real parts. So we must need

$$\beta_1 S^{**} - \beta_2(1-\alpha)(1-\eta)I_2^{**} - (\nu\eta + \mu + \mu_1) < 0,$$

Which corresponds to

$$R_2 > \frac{1}{\mu(1-\eta)}\left\{\frac{\beta_1(\mu+\mu_2)}{\beta_2(1-\alpha)} - (\eta\nu+\mu+\mu_1)\right\} + 1 = R^{**}. \,(say)$$

Hence, the equilibrium point E_2 is locally asymptotically stable provided the condition stated in the theorem is satisfied.

Theorem 4.4. The interior (coinfection) equilibrium point $E_3\ (\hat{S}, \hat{I}_1, \hat{I}_2, \hat{I}_3, \hat{I}_4)$ is unstable whenever it exists.

Proof: The variational matrix corresponding to coinfection equilibrium point $E_3\ (\hat{S}, \hat{I}_1, \hat{I}_2, \hat{I}_3, \hat{I}_4)$ is computed as follows:

$$M_3 = \begin{pmatrix} m_{11} & -\beta_1\hat{S}+\eta v & -\beta_2(1-\alpha)\hat{S} & 0 \\ \beta_1\hat{I}_1 & 0 & -\beta_2(1-\alpha)(1-\eta)\hat{I}_1 & 0 \\ \beta_2(1-\alpha)\hat{I} & -\beta_1\hat{I}_2 & 0 & 0 \\ 0 & m_{42} & \{\beta_1+\beta_2(1-\alpha)(1-\eta)\}\hat{I}_1 & -(\mu+\mu_3) \end{pmatrix}$$

where $m_{11} = -\{\beta_1\hat{I}_1 + \beta_2(1-\alpha)\,\hat{I}_2\} - \mu = 0$, $m_{42} = \{\beta_1 + \beta_2(1-\alpha_2)(1-\eta)\}\hat{I}_2$

Clearly, one eigenvalue of this matrix is given by $-(\mu+\mu_3)$ and other three eigenvalues are the roots of the following cubic equation which is the characteristic equation corresponding to principal minor of order 3 of the matrix M_3.

$$\psi^3 + A\,\psi^2 + B\,\psi + C = 0,$$

Here A = –(Sum of roots of this cubic equation), B = Sum of products of the roots taken in pairs and C = –(products of the roots). We know that product of the roots is the determinant of the matrix. Here it is easy to see that

$$C = -\beta_1\beta_2(1-\alpha)\,\hat{I}_1\hat{I}_2K < 0, \text{ where}$$

$$K = \left[\{\mu+\beta_1\hat{I}_1+\beta_2(1-\alpha)\hat{I}_2\}(1-\eta)+\{\beta_1-\beta_2(1-\alpha)(1-\eta)\}\hat{S}+\frac{v\eta\beta_2(1-\alpha)(1-\eta)}{\beta_1}\right] > 0,$$

as rate of TB transmission β_1 must be greater than rate of HIV transmission β_2. Hence the last cubic must have one root with positive sign, which implies that the variational matrix M_3 has at least one positive eigenvalue. Thus the coinfection equilibrium point is always unstable whenever it exists.

6.5 SIMULATION

The system (1) is simulated for various sets of parameters using XPP [13]. The stability of disease free equilibrium point E_0 is shown in Figures 6.2 and 6.3, where both the reproduction numbers R_1 and R_2 are less than 1 and the parameter values are as follows:

$$\Lambda = 30, \mu = 0.067, \beta_1 = 0.0006, \beta_2 = 0.0004, \mu_1 = 0.05, \ \mu_2 = 0.08,$$

$$\mu_3 = 0.09, \alpha = 0.35, v = 0.5, \eta = 0.4.$$

For this set of parameters, we get R_1 = 0.8474975, R_2 = 0.7919586 and the disease free equilibrium point E_0 (447.7612,0,0,0). The stability of TB-only equilibrium point is shown in Figures 6.4 and 6.5, when $R_1 > 1$ and $R_2 > 1$. Simulation is carried out for the following set of parameters:

$$\Lambda = 30, \mu = 0.067, \beta_1 = 0.001, \beta_2 = 0.0005, \mu_1 = 0.05, \mu_2 = 0.08,$$

$$\mu_3 = 0.09, \alpha = 0.35, v = 0.5, \eta = 0.4.$$

For this set of parameters R_1 = 1.412496 > 1 R_2 = 0.98994 < 1 and the equilibrium point E_1 is (317,74,0,0). The equilibrium points E_2 and E_3 do not exist for this set of parameters. Similarly for $R_1 > 1$ and $R_2 > 1$, we get HIV-only equilibrium point E_2, which is locally asymptotically stable. This fact is demonstrated in Figures 6.6 and 6.7 for the following set of parameters:

$$\Lambda = 30, \mu = 0.067, \beta_1 = 0.0007, \beta_2 = 0.0006, \mu_1 = 0.05, \mu_2 = 0.08,$$

$$\mu_3 = 0.09, \alpha = 0.35, v = 0.5, \eta = 0.4.$$

Here, R_1 = 0.988747, R_2 = 1.187938, E_2 = (376.923,0,32.2867,0. The equilibrium E_1 and E_3 do not exist and the disease free equilibrium E_0 (447.76,0,0,0) is unstable. When R_1 and R_2 both are greater than 1, we have possibility to visualize the existence and stability of the equilibrium point E_3, which is demonstrated in Figures 6.8 and 6.9. Figure 6.8 is corresponding to following set of parameters:

$$\Lambda = 40, \mu = 0.067, \beta_1 = 0.0008, \beta_2 = 0.0006, \mu_1 = 0.05, \mu_2 = 0.08,$$

$$\mu_3 = 0.09, \alpha = 0.35, \nu = 0.5, \eta = 0.4.$$

Here R_1 = 1.50666, R_2 = 1.5839, E_0 = (597.0149,0,0,0) E_1 = (396.25,114.9679,0,0), E_2 = (376.923,0,100.314,0) and E_3 = (414.1283,18. 1375,61.1224,7,3013). The constant term C of the quadratic equation (3.1) is –3.25 < 0. Figure 6.9 is demonstrating the same fact for the following set of parameters where C = 4.1 > 0:

$$\Lambda = 40, \mu = 0.067, \beta_1 = 0.0008, \beta_2 = 0.0006, \mu_1 = 0.05,$$

$$\mu_2 = 0.08, \mu_3 = 0.09, \alpha = 0.35, \nu = 0.6, \eta = 0.4.$$

It is verified that the local stability conditions for the stability of E_1 and E_2 are satisfied, so they are stable but the equilibrium points E_0 and E_3 are unstable. As the equilibria E_1 and E_2 are locally asymptotically stable under some conditions, so $R_1 > 1$ and $R_2 > 1$ do not guarantee the stability of these equilibria. This fact is demonstrated in Figure 6.10, where $R_1 > 1$ $R_2 > 1$, E_1 is stable, E_2 is unstable and E_3 = does not exist. Simulation is carried out for the following set of parameters:

$$\Lambda = 100, \mu = 0.067, \beta_1 = 0.00035, \beta_2 = 0.0003,$$

$$\mu_1 = 0.05, \mu_2 = 0.08, \mu_3 = 0.09, \alpha = 0.6, \nu = 0.5, \eta = 0.45.$$

Here R_1 = 1.52745, R_2 = 1.21839, E_0 = (1492.537,0,0,0), E_1 = (977.1429,295.1404,0,0), E_2 = (1225,0,121.9388,0). Figure 6.11 is demonstrating the same fact for the following set of parameters, where E_1 is unstable and E_2 is stable:

$$\Lambda = 30, \mu = 0.067, \beta_1 = 0.0012, \beta_2 = 0.001, \mu_1 = 0.05,$$

$$\mu_2 = 0.08, \mu_3 = 0.09, \alpha = 0.5, \nu = 0.6, \eta = 0.6.$$

For this set of parameters, we get $R_1 > 1.12644$, R_2 = 1.52299, E_0 = (447.7612,0,0,0), E_1 = (397.5,28.782,0,0), E_2 = (294,0,70.0816,0). To see the effect of screening of HIV infectives, we considered the case when both R_1 and R_2 are greater than one and both the equilibria E_1 and E_2 are locally asymptotically stable. It is observed that if our initial conditions are such that HIV only equilibrium is locally asymptotically stable then with the increase in the screening parameter α, first the equilibrium level of HIV infected population decreases and further increase in this parameter makes this equilibrium E_2 unstable and in this case TB-only equilibrium becomes stable. This fact is demonstrated in the Figures 6.12 and 6.13, where variation of HIV infected population I_2 and TB infected population I_1 with time are shown for different values of α. Similarly, the effect of screening of TB infected individuals is shown in Figures 6.14 and 6.15. It is observed that for the same initial start when we had local stability of TB only equilibrium point, increase in η causes instability of this equilibrium point and system tends to the HIV only equilibrium point. So we need to increase both the screening parameters η and α so that both the reproduction numbers R_1 and I_1 can be made less than one, which will cause disease free equilibrium to be stable, i.e. in this case system will tend to disease free equilibrium.

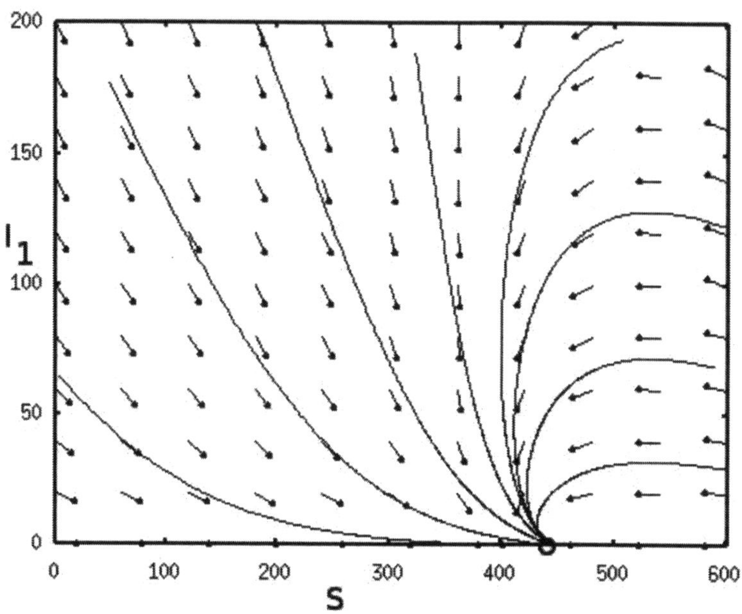

FIGURE 6.2 S– I_1 phase plane showing the stability of the disease free equilibrium point E_0.

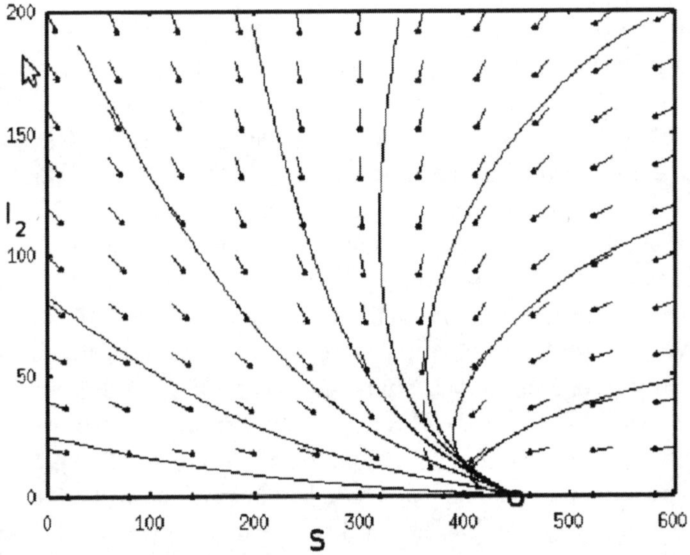

FIGURE 6.3 $S - I_2$ phase plane showing the stability of the disease free equilibrium point E_0.

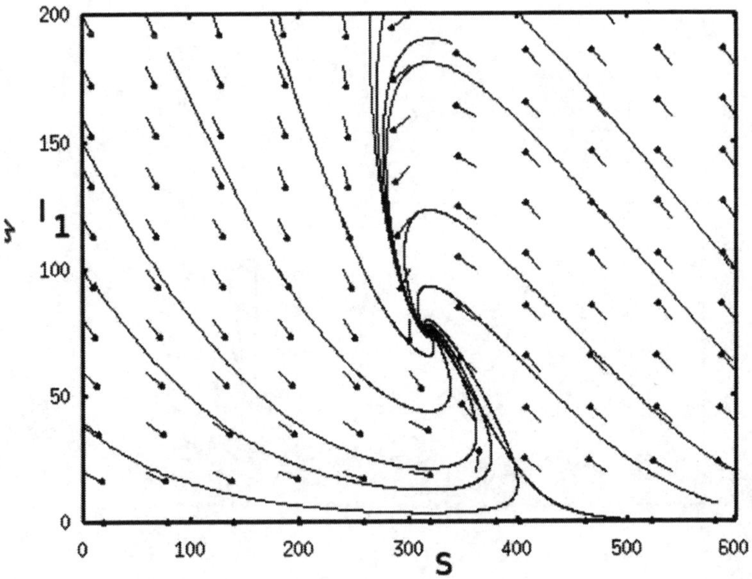

FIGURE 6.4 $S - I_1$ phase plane showing the stability of the TB-only equilibrium point E_1.

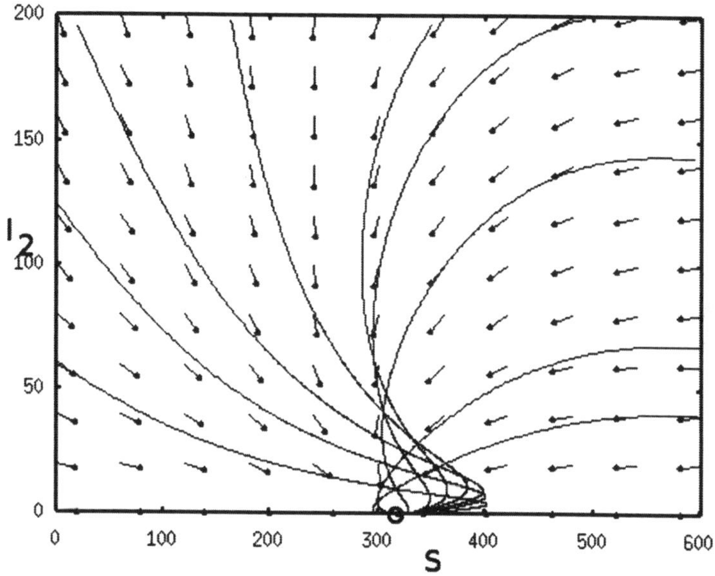

FIGURE 6.5 $S-I_2$ phase plane showing the stability of the TB-only equilibrium point E_1.

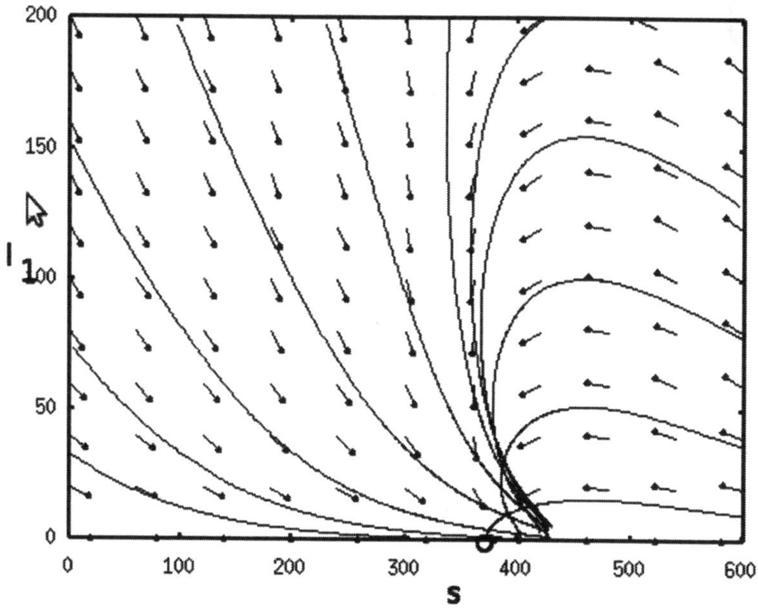

FIGURE 6.6 $S-I_1$ phase plane showing the stability of the HIV-only equilibrium point E_2.

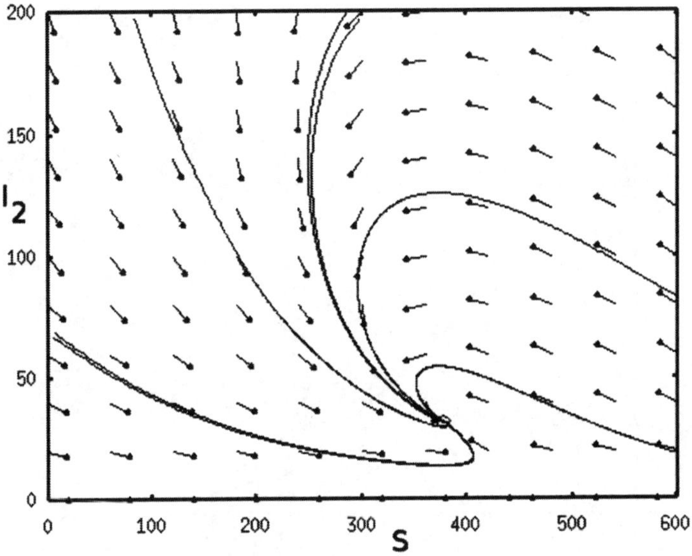

Figure 6.7 $S - I_2$ phase plane showing the stability of the HIV-only equilibrium point E_2.

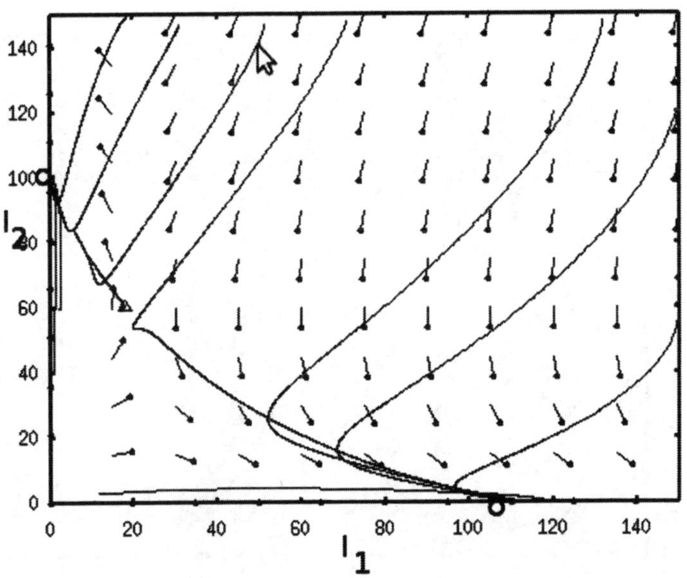

Figure 6.8 $I_1 - I_2$ phase plane showing the local stability of the equilibria E_1 E_2 and instability of the equilibrium point E_3 for $C < 0$.

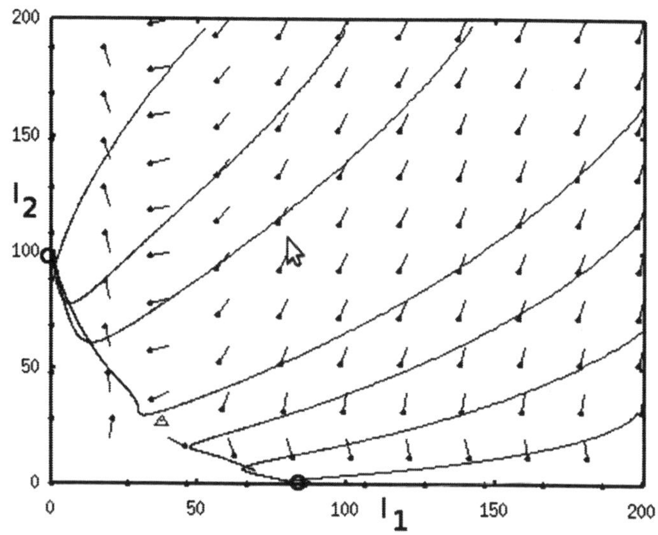

FIGURE 6.9 $I_1 - I_2$ phase plane showing the stability of the equilibria E_1 E_2 and instability of the equilibrium point E_3 for $C > 0$.

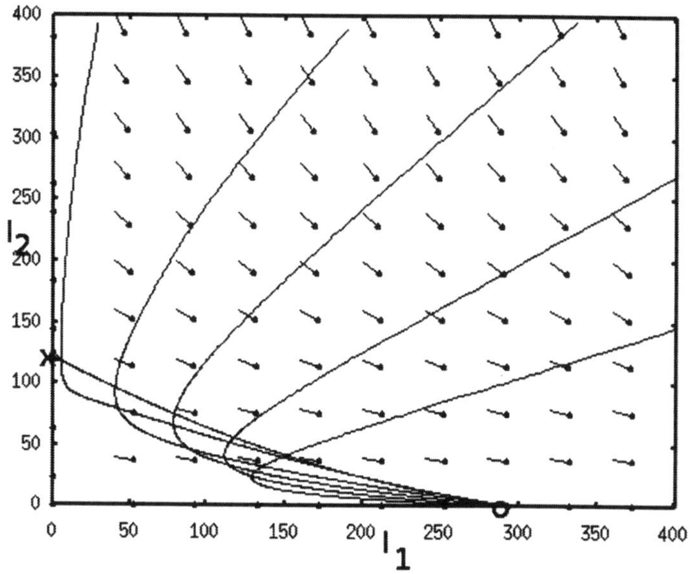

FIGURE 6.10 $I_1 - I_2$ phase plane showing the stability of E_1 and instability of E_0, E_2 when E_3 does not exist and $R_1 > 1$, $R_2 > 1$.

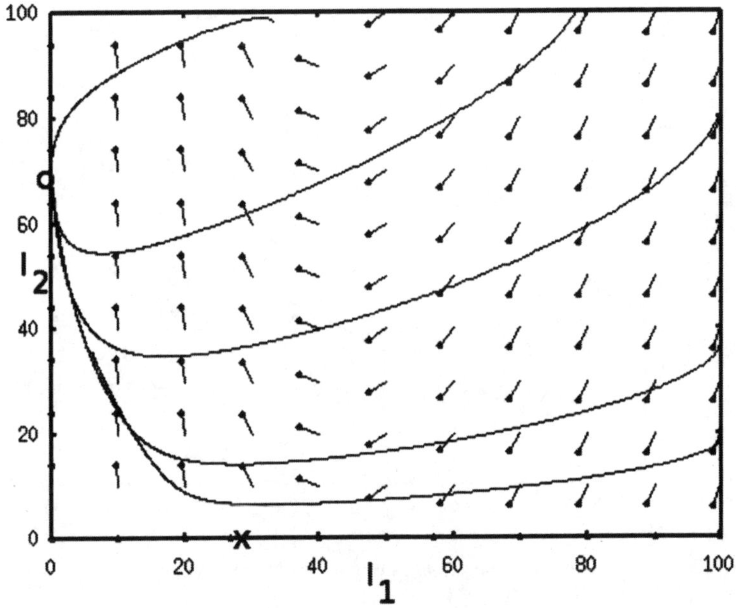

FIGURE 6.11 $I_1 - I_2$ phase plane showing the stability of E_2 and instability of E_1, E_0 when E_3 does not exist and $R_1 > 1$, $R_2 > 1$.

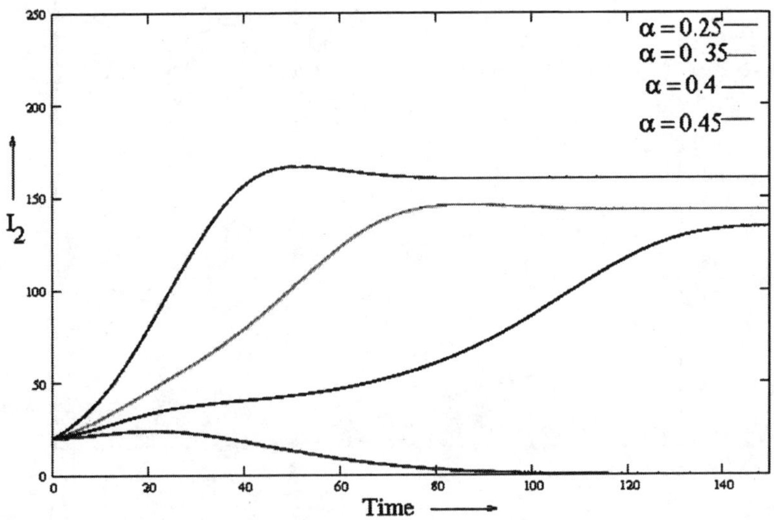

FIGURE 6.12 Variation of HIV infected population I_2 with time for different values of screening parameter α showing the effect of screening on the equilibrium level of HIV infected population.

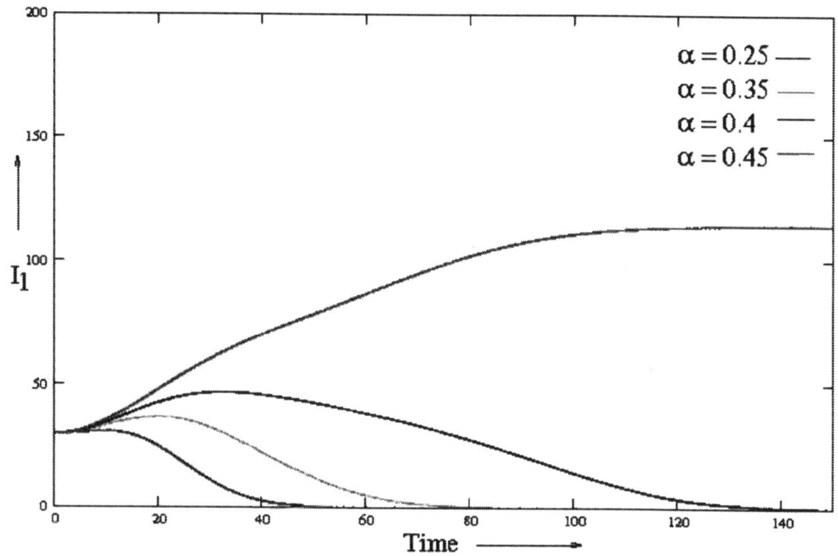

FIGURE 6.13 Variation of TB infected population I_1 with time for different values of screening parameter α showing the effect of screening on the equilibrium level of TB infected population.

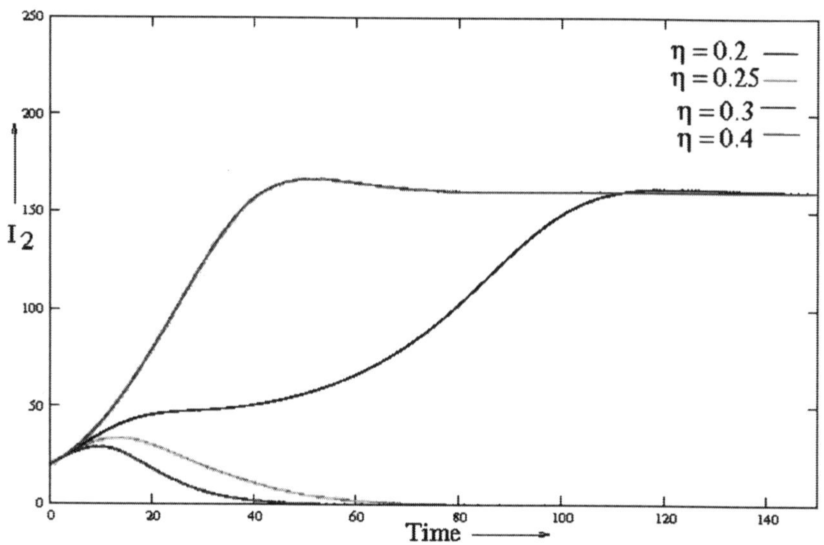

FIGURE 6.14 Variation of HIV infected population I_2 with time for different values of screening parameter η showing the effect of screening of TB infected individuals on the equilibrium level of HIV infected population.

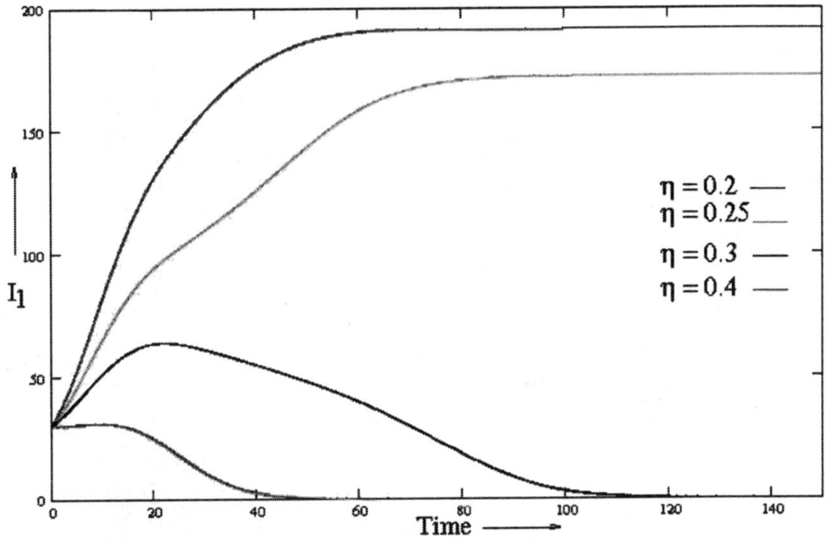

FIGURE 6.15 Variation of TB infected population I_1 with time for different values of screening parameter η showing the effect of screening of TB infected individuals on the equilibrium level of TB infected population.

6.6 CONCLUSION

In this work an HIV/TB co-infection model has been presented with the effect of screening of infectives. The basic reproduction numbers corresponding to both TB and HIV have been computed. The existence and stability of various equilibria have been discussed. We have shown that the disease free equilibrium is stable only when both the basic reproduction numbers are less than one. Furthermore, we have observed that the co-infection equilibrium is always unstable whenever it exists and we have either TB-only or HIV-only equilibria, which are locally asymptotically stable under certain restriction on parameters. Our numerical simulations have shown that the screening of infectives plays a very important role in the control of these diseases. Our results have shown that the increase in the parameters corresponding to screening of HIV and TB corresponds to the decrease in the equilibrium level of that infective population. But to eliminate both the diseases, we need to increase both the screening parameters so that basic reproduction numbers corresponding to both the diseases can be made less than one.

ACKNOWLEDGMENT

This research was supported by the research grants of DST, Govt. of India, via a sponsored research project: SR/S4/MS:681/10.

KEYWORDS

- **co-infection**
- **reproduction number**
- **simulation**
- **stability**

REFERENCES

Andrew Pollack, Donald G. Mcneil Jr., "In Medical First, a Baby with H.I.V. Is Deemed Cured", The New York Times, 3rd March 2013, website address: www.nytimes.com/2013/03/04/health/for-first-time-baby-cured-of-hiv-doctors-say.html?pagewanted=all&_r=0, 2013.

Ermentrout, B. Simulating, Analyzing, and Animating Dynamical Systems: A Guide to XPPAUT for Researchers and Students, 1st edition. SIAM, Philadelphia, PA, 2002.

Jun-Yuan Yang, Xiao-Yan Wang, Xue-Zhi Li, Feng-Qin Zhang, Souvik Bhattacharya, An HIV model: Theoretical analysis and experimental verification, *Computers and Mathematics with Applications*, 2011, 61, 2172–2176.

Lih-ing W. Roger, Zhilan Feng, Carlos Castillo-Chavez, Modeling TB and HIV Co-infections, *Mathematical Biosciences and Engineering* 2009, 6(4), 815–837.

May, R. M., Anderson, R. M. Transmission dynamics of HIV infection, *Nature* 1987, 326, 137–142.

May, R. M., Anderson, R. M., McLean, A. R. Possible demographic consequences of HIV/AIDS, *Math. Biosci.* 1988, 90, 475–506.

Oluwaseun Sharomi, Chandra N. Podder, Abba B. Gumel, Mathematical Analysis of the Transmission Dynamics of HIV/TB Coinfection in the Presence of Treatment, *Mathematical Biosciences and Engineering* 2008, 5(1), 145–174.

Ram Naresh, Agraj Tripathi, Dileep Sharma, A nonlinear AIDS epidemic model with screening and time delay, *Applied Mathematics and Computation* 2011, 217, 4416–4426.

Ram Naresh, Dileep Sharma, Agraj Tripathi, Modeling the effect of tuberculosis on the spread of HIV infection in a population with density-dependent birth and death rate, *Mathematical and Computer Modeling* 2009, 50, 1154–1166.

Srinivasa Rao, A. S. R. Mathematical modeling of AIDS epidemic in India, *Current Sci.* 2003, 84(9), 1192–1197.

UNAIDS, AIDS epidemic update, WHO, December 2007.

World Health Organization HIV/TB Facts 2011. http://www.who.int/hiv/topics/tb/hiv_tb_factsheet_june_2011.pdf (2011).

Yan, P. Impulsive SUI epidemic model for HIV/AIDS with chronological age and infection age, *J. Theoret. Biol.* 2010, 265(2), 177–184.

CHAPTER 7

STATUS AND TRENDS OF HIV AMONG HIGH-RISK GROUPS AND ANTENATAL WOMEN IN INDIA: ESTIMATES FROM SENTINEL SURVEILLANCE AND NFHS-3

SRINADH, PURNA CHANDRA DASH, and P. RAJENDRAN

CONTENTS

7.1 INTRODUCTION

HIV/AIDS assumed the status of a major public health problem in India demanding multi pronged public health interventions. Awareness and knowledge on HIV/AIDS among the people is a pre requisite for the success of the health interventions.

7.1.1 OBJECTIVES

This chapter is taken out from the study of Millennium Health Development Goals and India: A Review – 2007. The principal objective of the study is to review and assess the status and trends of HIV among high-risk groups and antenatal women in India and various issues and challenges that will be critical in achieving the goals.

7.1.2 METHODS

In the present study we have assessed the status of HIV/AIDS by using two sources of data, first we used the traditional approach of estimation by using the data collected from various sentinel sites. Next we used the NFHS-3 data for the same purpose.

7.1.3 RESULTS

The HIV infected adults (15–49) increased from 3.5 million to 3.97 million over the period from 1998 to 2001. Among the high prevalence states, the prevalence of STD and ANC is high in AP, Goa, Karnataka, Maharashtra and Manipur. Further, MSM and IDU are major factors contributing to the prevalence in Goa, Manipur and Nagaland and FSW contributes a chunk to the total prevalence in Maharashtra – particularly in Mumbai. NFHS-3 figures put the number of HIV-positive people in India at 2.5 million, almost half of previous estimates of 5.3 million. Further, it shows that the prevalence rate is also down to 0.36 from the earlier estimate of 0.91.

7.1.4 CONCLUSION

The Government of India established a National AIDS Control Program (NACP) to combat HIV epidemic. During NACP I (1992), there was increased awareness about HIV/AIDS, particularly among the urban population and subsequent successful intervention Programs and strengthening STDs clinics across the country were major achievements. During NACP II, the classification of states has focused on the vulnerability of states, with states being classified as high and moderate prevalence and high and moderate vulnerability. The primary goal of NACP III is to halt and reverse the epidemic by 2012.

7.2 BACKGROUND

Countries of the world started the new millennium in the year 2000 with an ambitious development agenda planned in the form of Millennium Development Goals (MDGs). The MDGs assume strategic importance for governments and development agencies as they provide a renewed focus and a way forward for the global development agenda. The Millennium Development Goals (MDGs) adopted by the United Nations in 2000, though not UN goals, represent a commitment by governments worldwide to do more to reduce poverty and hunger and to tackle issues of ill-health, gender inequality, lack of education, lack of access to clean water, and environmental degradation, besides commitments to reduce debt, increase technology transfers and build development partnerships, all by the year 2015. The MDGs were translated into targets that can be measured through verifiable and internationally comparable indicators. The first step to take forward the MDGs is to assess where the country stands in relation to MDGs and also review various issues and challenges that will be critical in achieving the goals.

In total, eight goals were planned:

1. Eradicate extreme poverty and hunger;
2. Achieve universal basic education;
3. Promote gender equality and empower women;
4. Reduce child mortality;
5. Improve maternal health;
6. Combat HIV/AIDS, malaria, and other diseases;
7. Ensure environmental sustainability; and
8. Develop a global partnership for development.

India was at the midway mark (2007) of achieving the Health MDGs, and therefore a critical review was necessary to take stock of current situation in order to make early corrections, if required. In earlier days (pre-2000) the communicable diseases in general and HIV/AIDS in particular, was showing very slow progress due to lack of knowledge about the use of contraceptives and their use among adults. For achieving goal 6 within the stipulated time period, it was necessary to undertake a critical review of all the secondary data available and make an analysis in order to assess the status and trends of HIV among core groups and ANCs.

This chapter is an extract taken from a larger evaluation study on "Millennium Health Development Goals and India: A Review—2007". This review was conducted by Centre for Management and Social Research (CMSR, Hyderabad), India on behalf of World Health Organization (WHO), India, Country Office. The study uses the secondary data that were collected through large-scale countrywide surveys and many research studies conducted during the past decade.

HIV/AIDS is assumed the status of a major public health problem in India demanding multi-pronged public health interventions. Awareness and knowledge on HIV/AIDS among the people is a pre requisite for the success of the health interventions. HIV/AIDS is one of the major public health problems in India demanding multi-pronged public health interventions. It is assumed that the interventions on increasing the awareness and knowledge on HIV/AIDS among the people in general and adults, in particular, would help mitigate the problem to a large extent.

7.2.1 OBJECTIVES

The primary objective of the paper is to review and assess the status and trends of HIV among High-risk Groups and Antenatal Women in India and various issues and challenges that will be critical in achieving the goals.

Specific objectives of the study are as follows:

- Updating the data pertaining to: status and trends of HIV among High-risk Groups (i.e., prevalence rate among IDUs, FSWs, MSMs, STDs, etc.) and antenatal women from recent surveys/sources.
- Examine the main differences of amongst high prevalence states/ group of states.

- Critical review of present policies and programmes, which impact on progress of the goal (NACP III).

7.2.2 METHODOLOGY

In the present chapter, we have assessed the status of HIV/AIDS by using two sources of data, though NFHS-3 estimates are the latest one. First we use the traditional approach of estimation by using the data collected from various sentinel sites. Next we use the NFHS-3 data for the same purpose. This gives us a scope for overall comparison of the quality of information as well as their reliability from these two sources of information.

The data for this study are largely drawn from the following surveys:

1. NACO, HSS-2007 (HIV Sentinel Surveillance); and
2. HSS 2010–11 provisional findings.

Though the reports of different surveys were published at later dates, the data used for this analysis pertains to the years 2003–2007 and presented for high prevalence states.

7.2.3 ANALYSIS OF DATA

1. HIV mean prevalence among different groups for selected high prevalence states.
2. HIV prevalence among ANC attendees and STD attendees for selected high prevalence states.
3. HIV prevalence among various HRGs, bridge population and ANCs by year wise.
4. Analysis of policies and programs across the country.
5. Implication, challenges and suggestions.

7.2.4 STUDY FINDINGS

The state wise prevalence was calculated from the projected population (methodology described in the foot note). As per the estimates, Andhra Pradesh, Karnataka, Manipur, Nagaland and Maharashtra are the high prevalence states. The prevalence of STD and ANC is consistently high in

AP, Goa, Karnataka, Maharashtra and Manipur. Taking these two factors together these states can clearly be included in high-risk states. However, among the high prevalence states, apart from ANC and STD; MSM and IDU are major factors contributing to the prevalence in Goa, Manipur and Nagaland and FSW contributes a chunk to the total prevalence in Maharashtra – particularly in Mumbai (Figure 7.1).

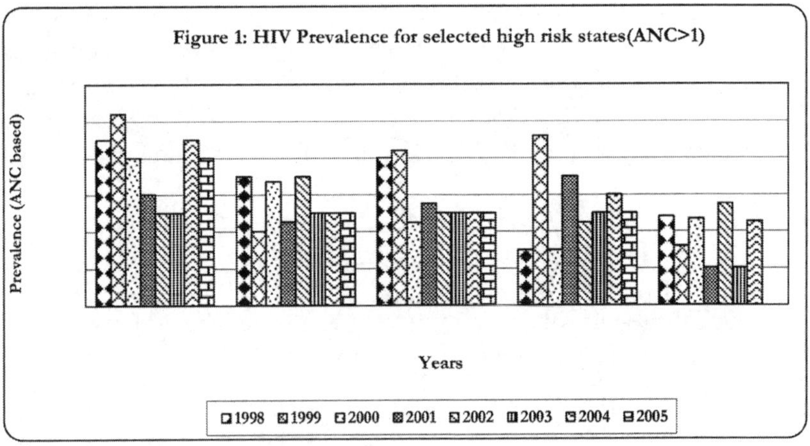

FIGURE 7.1 HIV Prevalence for selected high-risk states (ANC>1).

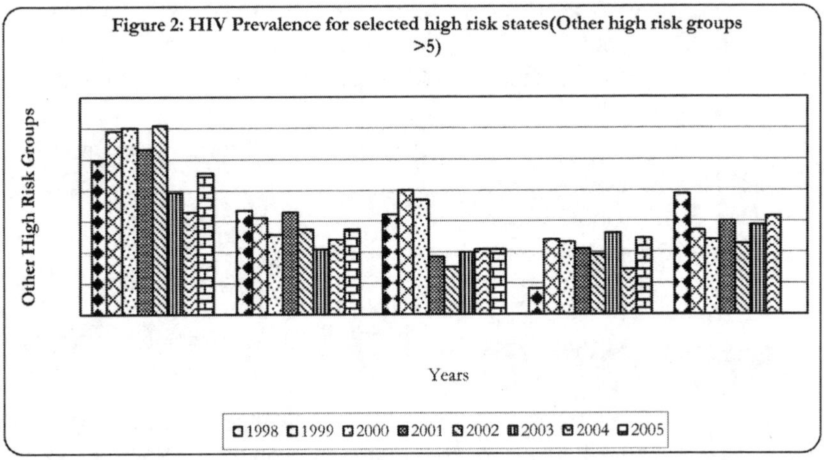

FIGURE 7.2 HIV Prevalence for selected high-risk states (Other high-risk groups >5).

Among the states which come under the moderate prevalence are Tamil Nadu, Delhi, and Mizoram. In almost all the states the prevalence of HIV among other high-risk groups is more than 5. In almost all these states the contribution of IDU is high. However, for Mizoram the incidence is gradually increasing and has crossed 1 (Figure 7.3) (NACO, HSS, 2006). Among the states which come under the moderate prevalence are Tamil Nadu, Delhi, and Mizoram. In almost all the states the prevalence of HIV among other high-risk groups is more than 5 (Figure 7.3).

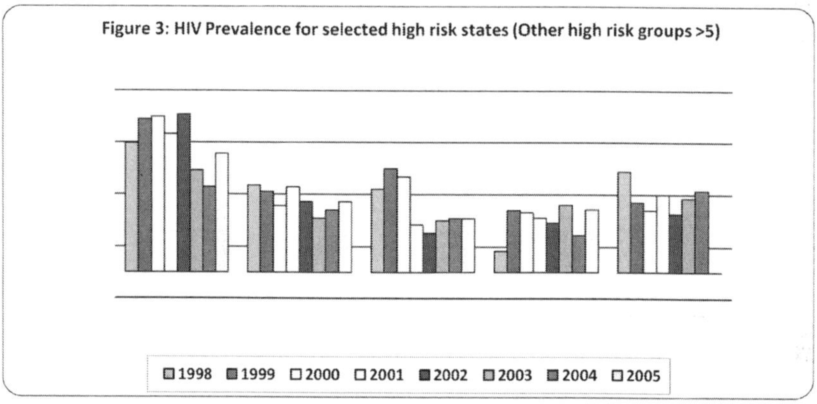

FIGURE 7.3 HIV Prevalence for selected high-risk states (Other high-risk groups >5).

Apart from these states, all other states are low prevalence states with ANC<1 and other high-risk groups<5.

Figure 7.4 depicts the HIV prevalence among HRGs and ANC prevalence (%) by year wise. The prevalence among high-risk groups is showing decreasing trend from 2003 to 2010, it is fluctuating among IDUs and MSMs. The HIV prevalence among ANC is slightly show increasing trend during 2003–05 and started decreasing since 2006, further it is remains constant during 2007–2010. The prevalence among truckers is showing decreasing trend during 2008 and it has increased in 2010. The prevalence among migrants has increased during 2007 and again decreased during 2008–2010. The prevalence among FSWs is also showing decreasing trend during 2003–2010.

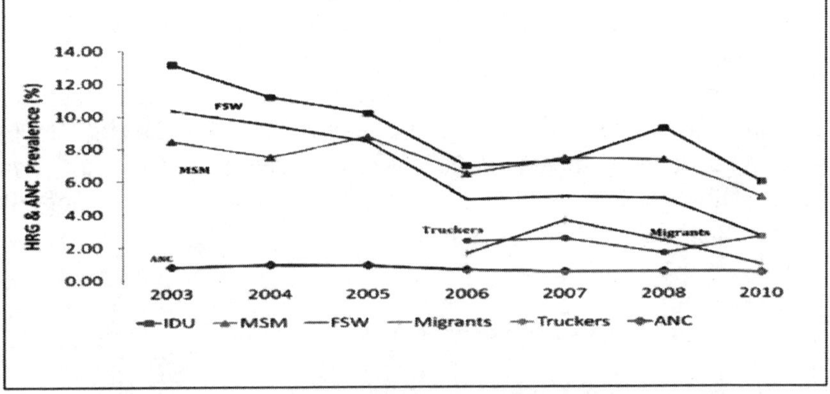

FIGURE 7.4 HIV Prevalence among HRGs and ANC Prevalence (%) year wise.

7.2.4.1 COMPARISON BETWEEN EAG AND NON-EAG STATES

There are 8 EAG states (Bihar, Jharkhand, Madhya Pradesh, Chhattisgarh, Uttar Pradesh, Uttaranchal, Orissa and Rajasthan). Almost all the EAG states are low prevalence states with ANC<1 and other high-risk groups<5. Though these states are not included in high or moderate prevalence states, it is observed that the growth rate of STD cases in Bihar, Chhattisgarh and Orissa is at a high pace over the years. It is therefore necessary that the interventions are targeted towards reducing the STD cases, specifically for these states. A comparative picture of HIV estimates by sex, rural-urban and among various sub groups of population is given in Table 7.1. The trend of infection among low, medium and high prevalence states is also reported in the same table.

TABLE 7.1 Comparative HIV Estimates in Various Sub-Populations (in millions)

Year	2000	2001	2002	2003	2004	2005
Total estimated HIV infection	3.86	3.97	4.58	5.10	5.13	5.21
Distribution by gender						
Infected males	1.94	2.04	2.58	3.22	3.13	3.18
	(60.2)	(61.5)	(68.14)	(63.06)	(61.1)	(61.0)
Infected females	1.24	1.24	1.21	1.89	2.00	2.03
	(39.8)	(38.5)	(31.86)	(36.94)	(38.9)	(39.0)

TABLE 7.1 *(Continued)*

Year	2000	2001	2002	2003	2004	2005
Urban rural distribution						
Infection in Urban areas	2.45	2.54	2.683	2.046	2.17	2.16
	(75.9)	(76.8)	(70.77)	(40.06)	(42.4)	(41.3)
Infection in rural areas	0.74	0.74	1.11	3.06	2.96	3.05
	(24.1)	(23.2)	(29.23)	(59.93)	(57.6)	(58.7)
HIV infection in various sub population groups						
Clients attending STD clinics	1.44	1.37	1.44	1.49	1.33	1.68
	(44.6)	(41.5)	(37.39)	(29.24)	(25.82)	(32.2)
Women attending ANC clinics (HIV infection among general population)	1.75	1.91	2.36	3.48	3.60	3.36
	(54.3)	(57.6)	(61.90)	(68.09)	(70.15)	(64.5)
IDU	0.04	0.03	0.03	0.01	0.01	0.01
	(1.1)	(0.9)	(0.71)	(0.20)	(0.21)	(0.19)
FSW				0.07	0.14	0.10
				(1.39)	(2.71)	(1.9)
New infection among children				0.05	0.06	0.06
				(1.08)	(1.11)	(1.2)
Population infection state wise						
Among high prevalent states	2.38	2.35	2.59	3.15	3.56	3.79
	(74)	(70.9)	(67.91)	(61.75)	(69.38)	(73.0)
Among medium prevalent states	1.26	1.38	0.11	0.18	0.11	0.12
	(3.9)	(4.2)	(2.88)	(3.43)	(2.05)	(2.3)
Among low prevalence states	0.71	0.83	1.12	1.78	1.47	1.29
	(22.1)	(24.8)	(29.21)	(34.82)	(28.57)	(24.7)
Population (Adult)	517.09	528.35	549.10	560.29	560.29	571.76
HIV Prevalence %	0.75	0.75	0.85	0.93	0.93	0.91

Source: NACO, HSS (2006).

Figures in the parenthesis are percentage to total.

7.3. PREVALENCE AS PER NFHS-3 ESTIMATES (2005–2006)

In the above section, we described the prevalence of HIV purely on the basis of sentinel surveillance data collected from the different sites. Because these surveillance estimates are not based on a representative sample of adults in India, it was decided to obtain general population estimates of HIV in India during NFHS-3. NFHS-3 is the first national survey in India to include HIV testing. NFHS-3 was designed to provide a national estimate of HIV in the household population of women age 15–49 and men age 15–54, as well as separate HIV estimates for each of the six highest HIV prevalence states (Andhra Pradesh, Karnataka, Maharashtra, Manipur, Nagaland, and Tamil Nadu). In addition HIV data from one of the low prevalent states (U.P.) was also collected to test reliability of the estimates.

7.3.1 STATUS IN THE COUNTRY

NFHS-3 figures put the number of HIV-positive people in India at 2.5 million, almost half of previous estimates of 5.3 million. Estimates of India's HIV caseload have been revised downward to about 2.47 million cases, less than half the number of cases estimated in previous studies. The new figures show that India has fewer HIV cases than South Africa and Nigeria; previous estimates pegged it as the country with the highest number of HIV infections.

The revised figures for 2006, based on data obtained through a household-based National Family Health Survey (NFHS-3), show that the prevalence rate (number of people infected per million) is also down to 0.36 from the earlier estimate of 0.91. This corresponds to an estimate of between 2 million and 3.1 million people living with HIV in the country.

The drastic drop in numbers comes from expanded surveys and improved methodology that provide a far more accurate picture of India's HIV status. The NFHS covered around 200,000 people between the ages of 15 and 54, and was conducted through face-to-face interviews across India between December 2005 and August 2006. AIDS experts from the United Nations, the World Health Organization, and the Bill and Melinda Gates Foundation have concurred with the new estimates.

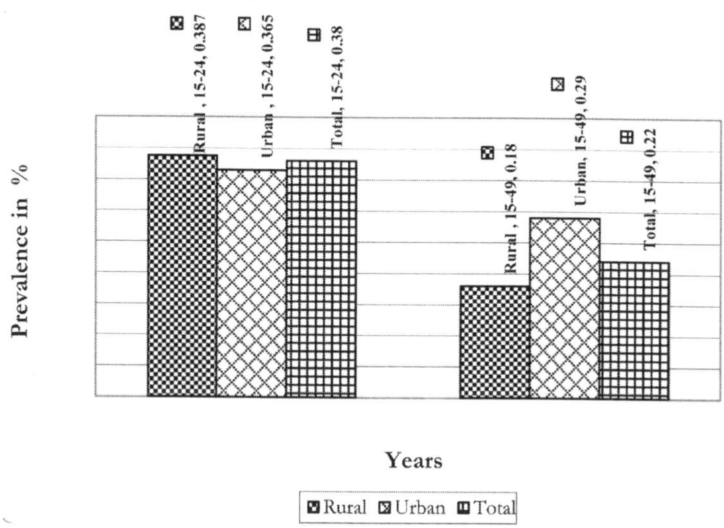

FIGURE 7.5 Prevalence of HIV Among women (15–24 and 15–49) – All India

7.4 PREVALENCE OF HIV AMONG 15–24 YEAR PREGNANT WOMEN

The prevalence of HIV among the age group 15–24 is shown in Figure 7.4. As per NFHS-3 estimates, the prevalence of HIV among this age group is 0.38. There is a slight difference between rural (0.387) and urban (0.365). In order to make a comparison between the sentinel surveillance estimates (Figure 7.1) and population based estimates (Figure 7.5), the prevalence figures for the age group 15–49 is also presented.

As per the population estimates, the prevalence of HIV between the age group of 15–49 was found to be 0.22, which is low compared to the estimates based on sentinel surveillance data.

7.5 CONCLUSIONS AND RECOMMENDATIONS

7.5.1 PRESENT POLICIES AND PROGRAMS

Shortly after reporting the first AIDS case in 1986, the Government of India established a National AIDS Control Program (NACP), which was managed by a small unit within the Ministry of Health and Family Welfare. India started combating HIV epidemic in 1990 through targeted IEC, establishment of surveillance system and blood safety interventions in few high prevalence states and vulnerable metros.

(a) **NACP I:** First nationwide control program was launched in 1992 (NACP-1) 1992–99. In 1992 the Government formulated a multi-sectoral strategy for the prevention and control of AIDS in India. It is implemented through the National AIDS Control Organization at the national level and State AIDS Cells at the State/UT levels. During this period there was increased awareness about HIV/AIDS, particularly among the urban which was almost insignificant at the beginning of the epidemic (Behavioural Surveillance Survey (BSS) 2000–01), subsequent successful intervention Programs and strengthening STDs clinics at various places of the country were one of the major achievements.
Initiatives and actions were taken to ensure blood safety by modernizing and strengthening the blood banks, introduction of licensing system for blood banks and gradual phasing out of professional blood donors.

(b) **NACP II:** Capacity and accountability at the state level continues to be a major issue and has required sustained support. Interventions need to be scaled up to cover a higher percentage of the population, and monitoring and evaluation need further strengthening. In order to induce a sense of urgency, the classification of states has focused on the vulnerability of states, with states being classified as high and moderate prevalence (on the basis of HIV prevalence among high-risk and general population groups) and high and moderate vulnerability (on the basis of demographic characteristics of the population).

(c) **NACP III:** Based on the experiences learnt from NACP-I and II, and in tune with the United Nation's Millennium Development Goal, India has now developed the third National AIDS Control Program (NACP-III, 2007–2012), with the primary goal to halt and reverse the HIV epidemic by 2012. The strategies will be prevention of new infections by saturating high-risk groups with targeted interventions, providing free ART to larger number of people living with HIV/AIDS, strengthening infrastructure and logistics support at all levels, and strengthening strategic information management system. The specific objective is to reduce new infections by 60 percent in high prevalence states and by 40 percent in vulnerable states in the first year. NACP-III target is to increase high-risk group's coverage from present 50 percent to 80 percent and to scale up ART coverage from less than 10 percent at present to 80 percent by 2011. It has been projected that if such coverage is achieved, number of people living with HIV/AIDS will come down from 5.7 million (2006) to 3.82 million by 2011, signaling a reversal of the epidemic.

NACP-III Program implementation will be more decentralized to make districts as the implementation unit, from where district specific interventions will be generated through formation of District AIDS Prevention and Control Units. Estimation, identification and mapping of high-risk groups at the district level will be one important strategy of NACP-III.

Under NACP-III, all existing counseling and testing centers will be remodeled as Integrated Counseling and Testing Centers (ICTC) will provide comprehensive services like BCC, condom promotion, treatment of STI, management of opportunistic infections, ART services, linkages with community support centers and also legal services. Number of centers will be increased from 2815 (2006) to 4955 (2011). Voluntary testing will be promoted vigorously to increase number of people tested from 6.5 million in 1st year to 22 million in the 5th year. Private sector will be encouraged to conduct similar number of tests. Improved access to safe blood will be ensured to bring down transfusion associated transmission, access to ART, referral for associated TB and the issue of drug resistance will be given special attention.

(d) Governmental and Community Based Organizations (NGOs & CBOs): There are numerous NGOs and CBOs working on HIV/AIDS issues in India at the local, state, and national levels. Projects include targeted interventions with high-risk groups; direct care of people living with HIV; general awareness campaigns; and care for children orphaned by AIDS. Funding for non-government and community-based groups comes from a variety of sources: the federal or state governments of India, international donors, and local contributions.

(e) **Donors:** India receives technical assistance and funding from a variety of UN partners and bilateral donors. Bilateral donors such as USAID, CIDA, and DFID have been involved since the early 1990s at the state level in a number of states. USAID has committed more than US$70 million since 1992, CIDA US$11 million, and DFID close to US$200 million. The number of major financers and the amount of funding available has increased significantly in the last year. Since 2004, the Bill and Melinda Gates Foundation has pledged US$200 million, and the Global Fund has approved US$54 million for HIV/AIDS for projects in rounds two, three and four. DFID (GBP 107 million) is providing pooled financing together with the Bank (US$ 250 million) in overall support to India's HIV/AIDS program NACP 3. Other donors include the Clinton Foundation, various UN agencies, DANIDA, SIDA, and the European Union.

7.5.2 IMPLICATIONS, CHALLENGES AND SUGGESTIONS

The facts given in Section 7.2 of this chapter gives enough evidence that India has been on the way of reducing the incidence and mortality due to HIV/AIDS. If the progress continues at the same pace, it is likely to achieve this MDG goal by 2015. However, so far as HIV/AIDS is concerned, there are several social taboos, which need to be addressed, and the present emphasis of the government and multilateral agencies should be towards designing the interventions that helps in reducing the social disbeliefs. The taboos (mentioned in section 1 of the chapter) mentioned earlier are to be looked into in detail and the budgetary allocations from the side of the government need to be enhanced further in order to address the issues. Utmost care must be taken to identify the high-risk groups and probably the targeted interventions will be one of the most effective methods to do so. As it is already known that the STIs are one of the major means of the transmission of the disease, the interventions should be targeted to the populations who are more prone to this infection.

Furthermore, appropriate arrangements from the side of the government need to be made to take appropriate care of the AIDS cases. This will certainly require research on clinical trials, which is much more expensive. As the epidemic is more among the adult population, including sex education in the school/college curriculum will be one of the cost effective method of handling the issue carefully. Since at the present context prevention is the only way to tackle the epidemic, appropriate ways are required for preventive as well as curative care for the epidemic, which needs appropriate surveillance system.

KEYWORDS

- **AIDS**
- **epidemic**
- **human immunodeficiency virus (HIV)**
- **National AIDS Control Program (NACP)**
- **National Family Health Survey (NFHS)**
- **population**

CHAPTER 8

CARE AND SUPPORT TO PEOPLE LIVING WITH HIV/AIDS (PLHIV) THROUGH COMMUNITY CARE CENTRE (CCC) APPROACH IN INDIA

HINDUSTAN LATEX FAMILY PLANNING PROMOTION TRUST (HLFPPT), INDIA

CONTENTS

ABSTRACT

Community Care Centre (CCC) is a comprehensive facility-providing medical, counseling, referral and outreach services to the registered PLHIV. The overall goal of the programme is to improve the survival and quality of life of people living with HIV/AIDS. Envisioned as a home away from home, CCC is a facility for providing accessible, affordable, and sustainable counseling, support, and treatment to PLHIV. With the medical services being an integral and important part of the programme, CCCs have a critical role in helping PLHIV gain easy access to ART treatment and counseling on primary prevention, nutrition, drug adherence, etc. The main functions of these CCC are providing treatment to the registered PLHIV on various types of Opportunistic infections/side effects, providing quality of counseling, providing home based care through outreach component and LFU tracking as per the list given by ART Centers. Currently 35 CCCs are in functional among 5 states and 1 union territory of India. A total of 66,471 PLHIV were registered with the CCCs and availing various services till December 2012 for smooth implementation of the programme, HLFPPT has been monitoring and providing constant technical support to the Community Care Centers (CCCs) scattered among the state above said states.

8.1 BACKGROUND

Hindustan Late Family Planning Promotion Trust (HLFPPT), is a Trust registered in 1992, promoted by HLL Life Care Limited, a Government of India Enterprise in the Ministry Health, for supporting implementation of public health program. With the vision of "Touching Lives with Quality Care, Compassion and emerge as a globally credible organization, HLFPPT's mission is to offer Innovative, Affordable and Sustainable Health Solutions. Over the years it has undertaken numerous pioneering projects especially in the field of HIV/AIDS that have helped it to empower and help many communities understand the basic necessity of a healthy lifestyle and thus contribute to the achievement of national health and population development goals. With experience of Pan India operations, HLFPPT is currently operational across 21 states working in the areas of Population Stabilization, RCH, HIV/AIDS and community empowerment.

8.2 NACP-III

The overall goals of NACP-III was to halt and reverse the epidemic in India over by integrating programmes for prevention, care and support and treatment through a four-pronged strategy:

- Prevent infections through saturation of coverage of high-risk groups with targeted interventions (TIs) and scaled up interventions in the general population.
- Provide greater care, support and treatment to larger number of PLHIV.
- Strengthen the infrastructure, systems and human resources in prevention, care, support and treatment programmes at district, state and national levels.
- Strengthen the nationwide Strategic Information Management System.

The specific objective is to reduce the rate of incidence by 60 percent in the first year of the programme in high prevalence states to obtain the reversal of the epidemic, and by 40 percent in the vulnerable states to stabilize the epidemic.

8.3 DESCRIPTION OF THE PACT PROGRAMME IN BRIEF

HLFPPT (as a sub-recipient (SR)) has been implementing Promoting Access to Care and Treatment Programme (PACT) in 5 states and 1 union territory of India with the support of Population Foundation of India (PR) and The Global Fund under GFATM. The implementing states includes Uttar Pradesh, Madhya Pradesh and Rajasthan, where the programme was initiated in 2008 and subsequently scaled up in Punjab, Chandigarh (UT) and Uttarakhand from April, 2012. The overall goal of the programme is to improve the survival and quality of life of people living with HIV/AIDS. For smooth implementation of the programme, HLFPPT has been monitoring and providing constant technical support to 35 Community Care Centers (CCCs), where the registered PLHIV have been receiving various types of services such as quality of treatment for minor OIs/side effects, quality of counseling and home based care through outreach workers to improve the survival and quality of life of PLHIV.

8.4 IMPLEMENTATION STRUCTURE OF PACT PROGRAMME

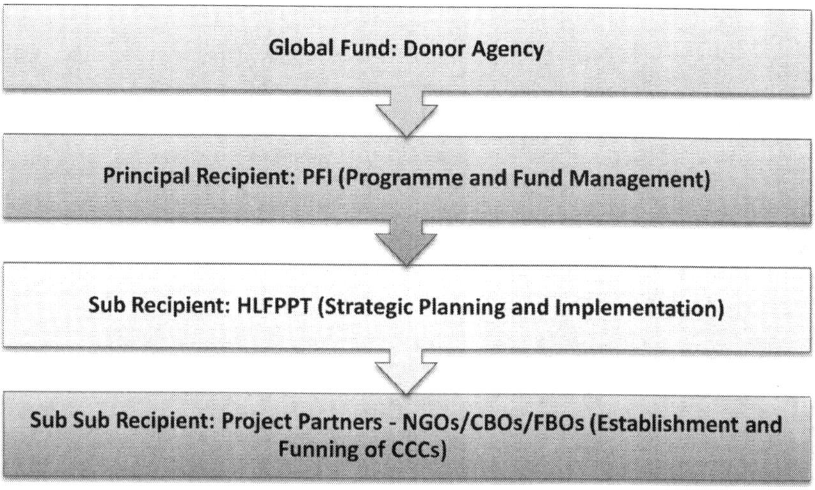

FIGURE 8.1 The Implementation Structure of PACT Programme.

8.5 GOAL, VISION AND OBJECTIVES OF PACT PROGRAMME

8.5.1 GOAL

To improve the survival and quality of life of people living with HIV/AIDS.

8.5.2 VISION

Improve access of PLHIV and the affected community with quality care, support and treatment services in line with National Guidelines.

8.5.3 OBJECTIVES

The main objective of the programme is to improve the survival and quality of life of people living with HIV/AIDS. The following are specific objectives to provide:

- in-patient and out-patient services;
- monitoring of side effects and tolerance to ART;
- Psycho-social and counseling support;
- treatment of opportunistic infections (OIs);
- nutritional support and counseling;
- referral services;
- outreach for follow-up of PLHIV for ART adherence;
- home based care and support;
- trace and retrieve defaulters (LFU & missed cases);
- accompanied referrals to the service delivery points;
- assistance PLHIV at ART centers;
- condom promotion.

8.6 MAJOR FUNCTIONS OF CCCS

The following are the major functions of CCCs:

- Providing treatment to the registered PLHIV on various types of opportunistic infections/side effects;
- Providing quality of counseling;
- Providing home based care through outreach component; and
- LFU tracking as per the list given by ART Centers.

8.7 SERVICE DELIVERY MODEL OF THE CCC

CCCs have been providing wide variety of services to registered PLHIV including medical and non-medical services. Medical services include treatment of side effects and other opportunistic infections/general ailments, which are normally appears in clients after using ART medicines and through the medical services clients gain strength to do their personal work and lead their families. Non-medical services include psychosocial support and counseling, home based care, advocacy against stigma and discrimination, nutritional support, referral to various Government Welfare Schemes, linkages to various income generation programmes, etc., which gives protection against to the social taboos and makes them economically stronger. The following services are included in the service delivery model of the CCC:

- Medical Services;
- Psychosocial Support and Counseling;
- ART Adherence;
- Education and Home Based Care;
- Education for Children;
- Advocacy against Stigma and Discrimination;
- Protection;
- Positive Prevention;
- Nutrition;
- Referrals and Linkages.

8.8 M&E SYSTEM OF PACT PROGRAMME

The monitoring and evaluation system is developed with an aim to:

- Collect evidence of activities and results,
- Assess the quantity and quality of achievements of targets, and
- Identify operational gaps in the programme and accordingly address them to improve programme efficiency.

The monitoring system focuses on the processes and outputs of the programme activities. The service delivery-based monitoring and evaluation (M&E) plan to support the scaling-up of care and support programme.

1. The M&E plan included minimal number of indicators that are easy to capture and analyze.
2. The indicators are consistent with earlier grant (Round 4) and thereby allowing more scope for comparison across regions.
3. The MIS was computerized at all levels starting from service delivery points and reports are shared with NACO/SACS periodically.
4. Operations research/special studies were conducted as appropriate.
5. The programme activity monitoring data is promptly analyzed, linked to routine supervisory system and fed back for programme improvement.
6. The MIS is built on the existing system and in line with the national strategic information system.

The MIS developed for the programme is according to the agreed activities and quarterly plans of each partner. The indicators are gender sensitive and measure progress over time and assess accomplishments against planned activities. The MIS is partner specific and thereby enables each one of them to measure their own progress. It encourages managers to identity operational gaps and thereby an opportunity to provide solution that ultimately increases the effectiveness of the programme.

8.9 MANAGEMENT INFORMATION SYSTEM (MIS)

The PFI has developed MIS for DLNs and CCCs under Round 4 programme. Keeping the same structure as in Round 4, both recording and reporting, the MIS of R4 programme is tailor made in order to meet specific requirements and activities committed in Round 6 grant. The information on age, sex and socio-economic status of each beneficiary was collected in addition to other individual and family history information. A total of 10 MIS registers and 12 MIS forms have been maintained under the programme to generate Monthly Quantitative Report.

8.10 COMPUTERIZED MANAGEMENT INFORMATION SYSTEM (CMIS)

The hearth of the routine monitoring system for PACT programme is the CMIS. Under R4 grant, PFI has developed this software for CCCs. The same is used after making necessary modifications as per plans for CCCs envisaged in R6 programme. For Community Care Centre, the software was developed. The CMIS is fed through a series of reporting formats that are compiled or used directly for data entry and to generate monthly or periodic (Quarterly) reports. This CMIS data enter on daily basis. Once uploaded into the system, data can be viewed at multiple levels of management and different types of analysis reports can be generated to aid in providing feedback and making programme decisions.

8.11 QUALITY ASSURANCE MECHANISM

The following system is followed to ensure validity and reliability of data:

- External MIS Audit – MIS audit is conceived as a mechanism for assessing the quality of programme data and assist partners to improve quality of their reporting system. It would essentially address reliability (at desk) and validity (authenticity of beneficiary's enrollment). This would involve assessing data quality and interaction with beneficiaries.
- This audit process starts at various levels of reporting. An independent external team will conduct this audit once in year 2. Subsequently, this will be carried out once a year.
- MIS Software – inbuilt validations: all fields are entered; valid data ranges and ensure consistency. In spite MIS in place, the SDPs were maintained the manual system too (like registers and other forms). The software allows quick yet quality reporting.
- The Project Coordinator of CCC also conducts spot checks to ensure the data first written on data collection formats translates to the correct value in the CMIS. In addition to regular supervisory review, occasional site visits by PR/SRs should include a record review to ensure that hard-copy registers and electronic reports show consistent counts.

The monitoring and supervision of PACT programme is being done at all levels from the donor to SR including NACO. Each agency has their set of indicators to monitor the programme on a monthly/quarterly basis. All the indicators in MPR are similar with all the agencies (Global Fund/ PR/SR/NACO). HLFPPT has its own monitoring system to monitor the programme on a regular basis (monthly/quarterly/need based) through monitoring tools, regular review meetings at all levels and on site visits to the CCCs.

8.12 REPORTING SYSTEM IN PACT PROGRAMME

Quarterly Progress Reports (QPRs) have been submitting on time to Global Fund. QPRs and MPRs have all the indicators of the programme including details of OP/IP, In/Out referral, counseling services, OIs treatment, staff training details, LFU cases tracked, outreach for home based care, etc.

The diagram of M&E system for PACT programme is given in Figure 8.2.

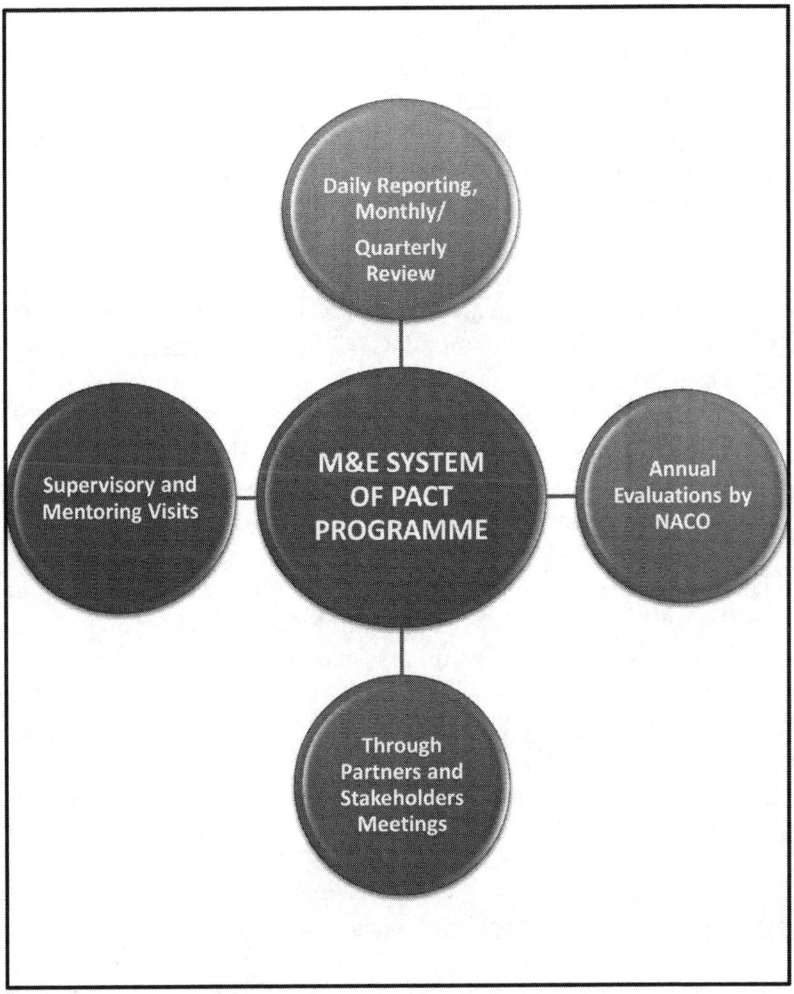

FIGURE 8.2 Diagram of M&E system for programme PACT.

8.13 PROGRAMME ACCOMPLISHMENTS (2008–2012)

8.13.1 CLIENT REGISTRATIONS

A total of 68645 clients were provided care and support in 35 CCCs (scattered across five states and one union territory) against the target of 37033, which shows nearly half of the clients were provided care and support additionally against the target during 2008–2012. Figure 8.3 depicts the achievement against the target in client registrations.

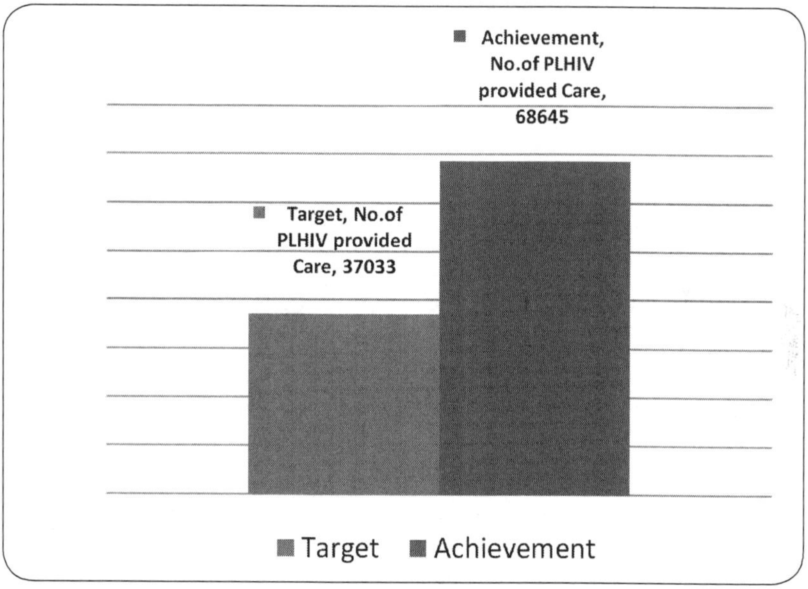

FIGURE 8.3 PLHIV provided care and support.

8.13.2 BED OCCUPANCY RATE (BOR)

The Bed Occupancy Rate gives an idea about the utilization of beds in the CCCs by registered clients. This is one of the key performance indicators in the programme. As per the Figure 8.4 except Punjab in all other states the BOR is between 60–90% and it is considered to be satisfactory as per the National Guidelines.

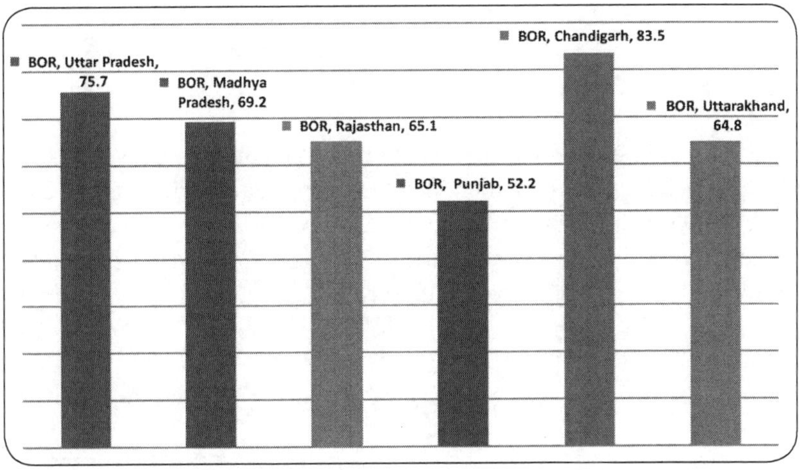

FIGURE 8.4 Bed occupancy rate.

8.13.3. BED TURNOVER RATIO (BTR)

The Bed Turnover Ratio shows the average length of stay by the registered clients in the CCCs. This is also one of the key performance indicators like BOR. As per the Figure 8.5 in all other states the BTR is 4 and it is considered to be very good as per the National Guidelines.

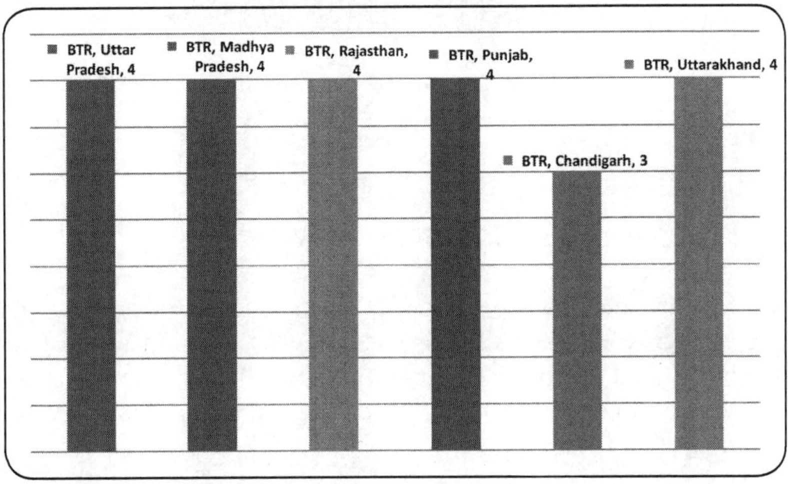

FIGURE 8.5 Bed turnover ratio.

8.13.4 DRUG ADHERENCE RATE (DAR)

Drug Adherence Rate gives the percentage of registered clients, who are adhered to ART medicines regularly and whose adherence level is more than 95% as per the National guidelines. All the CCCs capture the information on DAR of clients registered in CCCs from nearby ART centers and it is the responsibility of CCCs to track DAR on a monthly/quarterly basis. As per the Figure 8.6, except Chandigarh in all the states the drug adherence rate is above 90 percent.

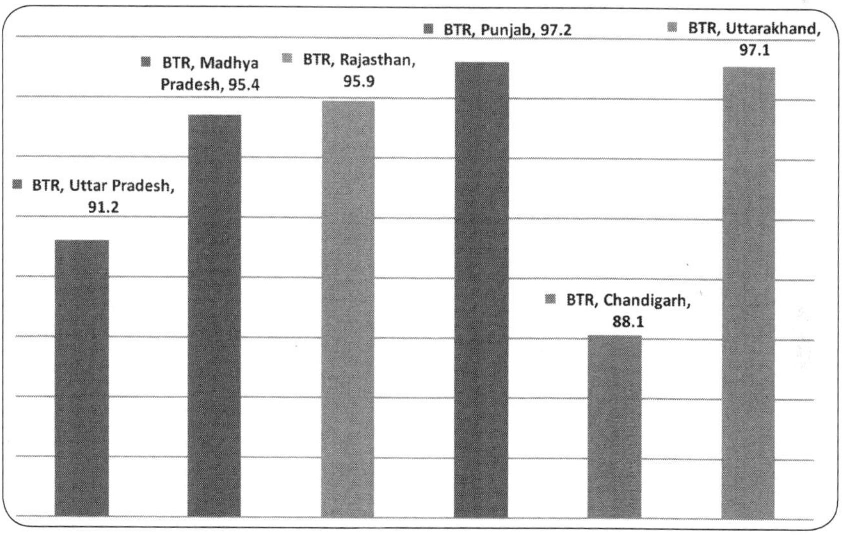

FIGURE 8.6 Drug adherence rate.

8.14 TREATMENT OF VARIOUS OPPORTUNISTIC INFECTIONS

It is one of the key functions of CCCs to give treatment to the registered clients, who are prone to various OIs. CCCs are limited to treat minor OIs like Tuberculosis, Candidiasis, Diarrhea, Bacterial Infections, etc. Figure 8.7 depicts state wise treatment of various OI episodes.

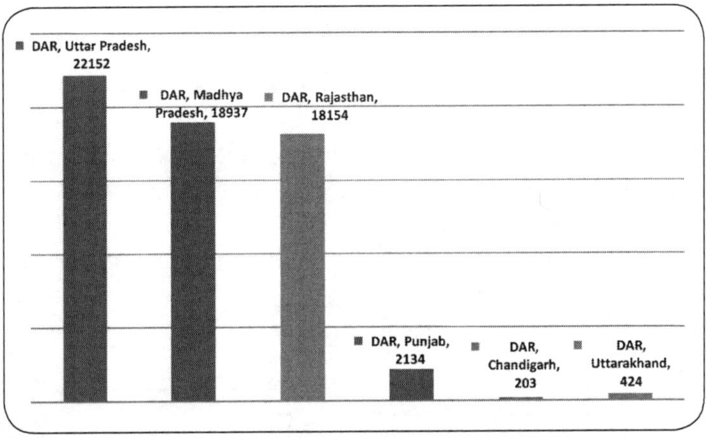

FIGURE 8.7 Treatment of various OIs (Episodes).

8.15 COUNSELING SERVICES

Providing psychosocial support to the registered clients plays a key role in changing their behavior and attitude towards the epidemic and it enables them to create a healthy environment within them and among their families. All the registered clients were received quality of counseling services by the professional counselors in the CCCs. The counselor provides counseling on drug adherence, which is considered as one of the most important counseling service to the on ART clients, who has to take ART medicines regularly and counseling other issues such as positive living, OIs and side effects, basics of HIV, sexual and reproductive issues, etc. Figure 8.8 depicts the state wise number of PLHIV received counseling services.

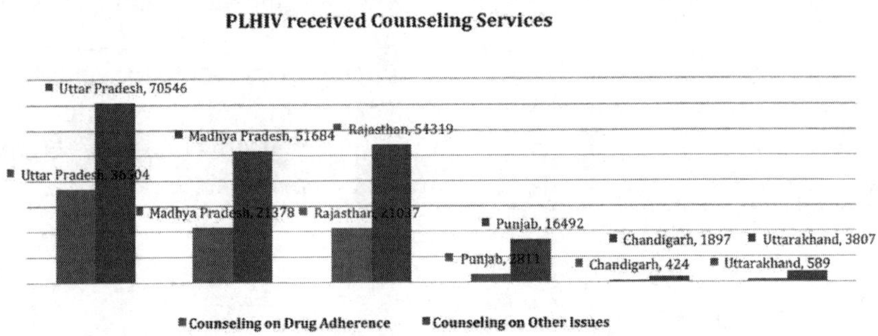

FIGURE 8.8 PLHIV received counseling services.

8.16 OUTREACH ACTIVITIES

Providing a home based care to the clients plays a key role in the lives of PLHIV, it touches their hearth and improves their quality of living. CCCs were successfully implemented outreach activities through outreach workers. Figure 8.9 depicts the state wise home visits made by outreach workers and the reasons for home visits include verification of address, follow-up for CD4 test, home based care, referrals for medical services, ART adherence and assessment for counseling and other services as required, etc.

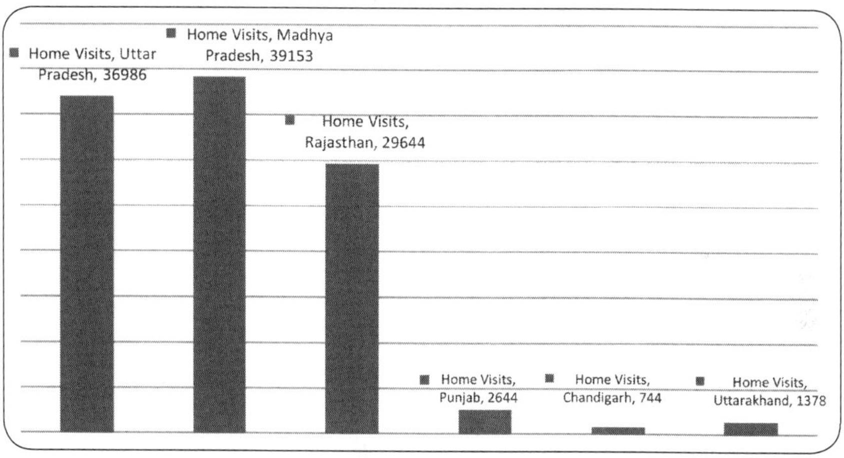

FIGURE 8.9 PLHIV contacted at their home.

8.17 REFERRAL TO GOVERNMENT SOCIAL WELFARE SCHEMES

It is one of the major responsibilities of CCCs to link their clients to various Government Social Welfare Schemes in order to improve their economic status and quality of life. Through the linkage to various schemes, clients enable to avail the schemes implemented by various Government Welfare Departments and getting benefitted. Figure 8.10 depicts the number of referrals made by the CCCs to the Government Social Welfare Schemes.

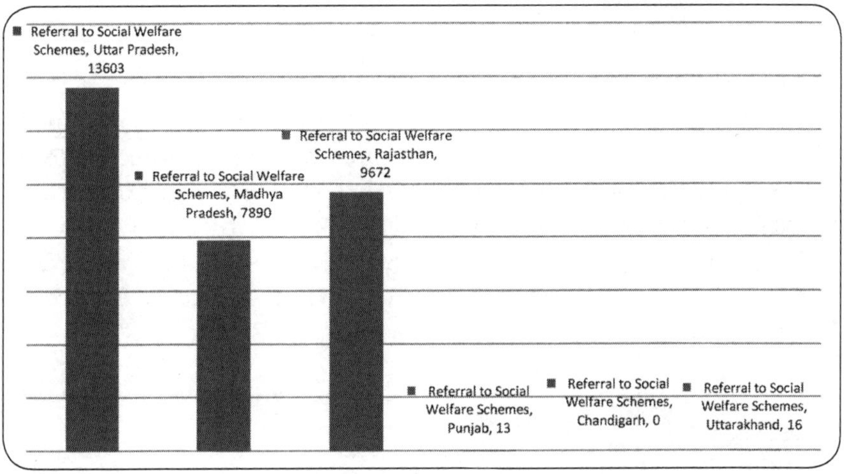

FIGURE 8.10 PLHIV Referred to Govt. Social Welfare Schemes.

8.18 BEST PRACTICES IN THE PROGRAMME

The Best Practices adopted by HLFPPT under the PACT program are:

- Continued Comprehensive Education (CCE);
- Strong outreach strategies were developed and were implemented which complimented to the project;
- Development of phone directory by the counselors;
- Using local dialect while interacting with the PLHIV, so that they gain a good understanding;
- Sharing of monthly outreach plan with the ART centers;
- Monthly coordination meetings with the ART center at Meerut;
- Yoga classes for the mind, body and soul at the CCC centers;
- Group counseling – to make the clients understand that there are others too with similar condition and to gain from similar experiences;
- Introduction of address verification certificate;
- Introduction of health and hygiene maintenance chart;
- Introducing the solar heating system at CCC II Indore;
- Introduction of the supervisory tool;
- Introducing the drug procurement protocol;

- Observance of special days like HIV/AIDS day, Deepawali, Christmas, Rakhi and other events to make the clients feel at home.

8.19 INNOVATIONS IN THE PROGRAMME

The following are the innovations made during the implementation of the programme by HLFPPT:

- Partners Meet
- Onsite Training
- Monitoring Visits Using Supervision Tool
- Continued Comprehensive Education (CCE)
- Outreach Strategy
- Financial Guidelines for CCC
- Drug Procurement Protocol

8.20 PROGRAMME CHALLENGES

The following are the programme challenges observed during the implementation of the programme by HLFPPT:

- Initial challenge was to set up the Community Care Centre;
- Identification of dedicated partner organization to implement the CCC;
- Coordination among various stakeholder with SACS, DLN, ART, TB DOTS and working for a common cause;
- Maintaining adherence to treatment with ARV;
- Identifying defaulters (LFU/missed cases) and linking back to treatment;
- Facilitating livelihood program is a major challenge in the programme;
- Identification and retention of competent staff at the CCCs within the range of salary;
- Maintenance of quality services despite turnover of staff.

8.21 LEARNING FROM THE PROGRAMME

PACT Programme has given a lot of programmatic and operational learnings. The following are the key learnings from the programme:

- After working with the NGOs, FBOs & CBOs, it is observed that social and professional commitment of the Partners determines the quality of the services at CCC.
- It is experienced that the systems and procedures would strengthen the process of sustainability of the programme in a long run.
- Most of the clients accessing services from CCCs belong to the lower income groups. Innovative approach to support them with livelihood options would enhance the quality life and medical needs of the PLHIV.
- Strong Coordination needed among the referral points to address HIV/AIDS in a holistic manner. Coordinated approach is required for psycho-social development of the PLHIV. This is enhanced through coordination efforts in establishing and strengthening linkages and referral mechanism.
- Budgetary constraints are directly affecting the outreach activities as it is limited to the district, where CCC is operational.
- Turnover of the trained staff is another challenge faced under the programme. This is frequent among nursing staff as well as the counselors.

8.22 SUMMARY AND CONCLUSIONS

Overall, the programme has been implemented successfully by HLFPPT with the support of PFI and Global Fund during 2008–2012. All the activities planned under the programme to implement are duly accomplished in time. The key activities in the programme including enrollment of all new cases ART (priority), follow-up of the LFU/missed cases, comprehensive Counseling to all (IP & OP) and reinforcement through the out-reach visits, maintenance of Drug adherence of all the registered cases, OI treatment to all the needy, referral and linkages strengthened, monitoring, Involvement of NGOs/FBOs/CBOs in monitoring, documentation and timely reporting, linkage with Government schemes and other development programmes like transport, insurance, nutrition and livelihood, etc., influencing social support and encourage public participation, strengthen community to handle social responsibility, strengthening coordination and partnership for promoting livelihood & income generation initiatives for the PLHIV.

All the stakeholders in the programme such as DLNs, ART Centers, SACS other agencies are well cooperated and provided their constant and hand holding support to the programme staff as and when required. All the ART Centers played a key role in referring the needy clients in time to the CCCs. Clients have received quality of services in CCCs including medical and non-medical services during their stay. Outreach component played a key role in providing the home based care and reinforcement of messages given at CCC level.

KEYWORDS

- **anti retroviral therapy**
- **community-based organization**
- **community care centre**
- **computerized management information system**
- **direct observation treatment system**
- **district-level network**
- **faith-based organization**
- **loss to follow-up**
- **management information system**
- **National AIDS Control Programme**
- **opportunistic infections**
- **Promoting Access to Care and Treatment (PACT)**

REFERENCES

Annual Reports of PACT (2008–2012).
NACP-III Guidelines 2007.
PCT CMIS Quantitative Reports (2008–2012).
Quarterly Reports of PACT.

CHAPTER 9

OPPORTUNISTIC INFECTIONS IN HIV/AIDS: AN OVERVIEW

BEENA ANTONY

CONTENTS

9.1 INTRODUCTION

The greatest challenge to public health in the recent years is the emergence and pandemic spread of HIV/AIDS. Since the first documented case of HIV in 1981, more than 60 million cases and 30 million deaths were reported till-to-date. Approximately, 2.7 million people are newly infected in 2010, which accounts to more than 7,000 cases everyday. A rapid rise in plasma viraemia coupled with a drop in CD4 count is the characteristic feature of acute HIV infection. Due to the progressive decline in the immunological response, HIV infected persons become a 'microbial zoo.'

According to the recent Annual report released from the National AIDS Control Organization (NACO), the estimated adult HIV prevalence in India was 0.32% and the total number of people living with HIV was 24 lakhs in 2009, 83% of this population belong to the age group 15–49 years. The first National AIDS Control programme (NACP) launched in 1992, followed by NACP II in 1999 and NACP III in 2007 were effective in the prevention and control of HIV/AIDS in India. Analysis of these HIV Surveillance programme revealed that the estimated annual HIV incidence has declined by about 56% over the last decade (2000–2009) from 2.7 lakh new infections in 2000 to 1.2 lakh in 2009. Among the Indian states, Manipur has shown the highest estimated adult HIV prevalence (1.4%) followed by Andhra Pradesh (0.9%). It is also observed that some low prevalence states have shown slight increase in the number of new infections over the past two years, which underscores the need for the programme to focus on these states.

Patients with HIV infections are extremely susceptible to a variety of opportunistic infections (OI). Major cause of morbidity and mortality of these patients are due to OI. Clinical profile and pattern of OI vary from patient to patient and from region to region. In HIV/AIDS, organisms of low virulence become opportunistic pathogens and known pathogens can be present in an abnormal fashion with increased virulence as well as with atypical clinical presentation. Early effective diagnosis depends upon high index of clinical suspicion followed by good laboratory practice. Identifying the causative agent in an earlier phase is essential in the early management of the case. Symptoms associated with the initial stage of the infection have been found to be fever, sore throat, skin rash, lymphadenopathy, splenomegaly, myalgia, arthritis and meningitis with varying degree of severity. The acute phase is followed by a latent phase without any clinical events, which may last for months to years characterized by low viral replication and gradual fall in CD4 count.

The awareness and impact of HIV/AIDS is different in developed countries and developing countries. Various studies show that about 75% of infected people are aware of their HIV sero status in the USA and Europe, while only 10–20% in Indians. There are many factors attributing which include limited access to health care facilities by the population, lack of basic infrastructure for early diagnosis of HIV, which requires well-established laboratories, lack of funds to support awareness or preventive campaigns as well as the specific treatment like highly active antiretroviral therapy (HAART). As reported by certain investigators, the spectrum of OI and the prevailing HIV sero-status are expected to change as a result of HAART. Hence an early detection and diagnosis of OI may help in effective disease management.

In this chapter, an attempt is made to review the various important opportunistic infections encountered in patients with HIV/AIDS with an emphasis on diagnostic and therapeutic approach. Spectrum of commonly encountered Opportunistic infections in HIV/AIDS is given in the Table 9.1.

TABLE 9.1 Opportunistic Infections in HIV

Bacterial Infections	**Mycotic Infections**
M.tuberculosis	Candidia
Atypical Mycobacteria	Cryptococcus
Salmonella species	Aspergillus
Campylobacter	Pneumocystis jerovecii
Legionella pneumophila	Histoplasma capsulatum
	Penicillium marneffei
Viral Infections:	**Parasitic infections**
Herpes,	Toxoplasma gondii
CMV	Cryptosporidium parvum
Malignancies:	Isospora belli
Kaposi's sarcoma	Strongyloides stercoralis
Hodgkin's Lymphoma	Microsporidia
Squamous cell carcinoma	

9.2 BACTERIAL OI

Among the bacterial OI, tuberculosis was the commonest one as reported in few studies from India. Tuberculosis is endemic in India and is considered as the most common cause of death in AIDS patients. Relative rarity of cavitary lesions and the confusion in distinguishing radiographic infiltrates of tuberculosis and other pulmonary pathogens are the main problems encountered in the diagnosis. Hence high index of suspicion and availability of specialized laboratory techniques such as microscopy, which demonstrates acid-fast bacilli and culture of mycobacteria on Lowenstein Jensen's medium are required.

Extra pulmonary tubercular manifestations with multi organ involvement occur in 46–79% of patients with pulmonary TB and HIV. As tuberculin positivity is less prevalent among HIV seropositive patients, Mantoux skin testing for evaluating tuberculosis is also a challenging task. HIV infected patients with active tuberculosis were found to have higher plasma viral loads compared to HIV infected persons with other OIs.

Atypical Mycobacteria constituting Avium-intracellularae complex is the next predominant bacterial pathogen of immunocompromised patients. It affects multiple organs and symptoms are similar to M.tuberculosis, but with less severity. This group of organisms is more frequently reported in the developed countries.

Community acquired **bacterial pneumonia** due to *S.pneumoniae* and *H.influenzae* was reported as another common OI in HIV infected patients with two-fold higher incidence than the general community. Bacterial pneumonia is reported to occur relatively early in the course of HIV. Disseminated pneumococcal infections are more frequent in individuals with low CD4 count.

Salmonellosis is emerged as another significant OI in HIV /AIDS. In a study involving 3000 hospitalized African patients, incidence of Salmonella infection was 13% in HIV positive patients, while 5.4% in HIV negative patients. Among these isolates, *S.Typhimurium* and *S.Enteritidis* accounted to 41% and 26%, respectively. A higher incidence of bacteraemia and focal infections were reported in these cases, which led to relapses in 12% and death in 24%. Non Typhoidal Salmonellae infections associated with HIV were found to be twenty times more compared to the general population.

9.3 PARASITIC OPPORTUNISTIC INFECTIONS

Gastrointestinal infections are very common in patients with HIV infection or AIDS with diarrhoea being a common clinical presentation. Diarrhoea is reported to occur in 30–60% of AIDS patients in developed countries and in about 90% of AIDS patients in developing countries with varying aetiological spectrum, which includes bacteria, parasites, fungi and viruses. Cryptosporidium parvum, Cyclospora cayetanensis, Isospora belli and Microsporidia are significant **opportunistic parasites,** while Entamoeba histolytica, Giardia lamblia, Trichuris trichiura, Ascaria lumbricoides, Strongyloides stercoralis and Ancylostoma duodenale are designated as **non-opportunistic parasites** in these patients. Opportunistic parasitic infection in immunocompromised is accompanied by chronic diarrhea and weight loss.

C.parvum, *I.belli* and *E.histolytica* were reported as the most frequently encountered organisms in HIV infected individuals with diarrhea, from India and other parts of the world. It is also observed that the proportion of the opportunistic pathogens in patients with advanced disease, i.e., CD4 count <200 cells was significantly higher as compared with patients with CD4 count >200 cells. Although cryptosporidiosis can be acquired at any time during the course of HIV infection, major morbidity and mortality occur almost exclusively in patients with CD4 counts below 200 cells. Spontaneous recovery was reported to occur above this level. Laboratory diagnosis of Cryptosporidium is very simple as it can be observed as acid-fast oocytes by modified acid fast stain.

In patients with AIDS, **Cryptosporidium** occur as a permanent diarrhoeal illness, which leads to chronic malabsorption syndrome. It is reported that in patients who fail to respond to therapy, death usually occurs in 3–6 months. As chlorination is not effective in preventing cryptosporidiosis, the potential for waterborne transmission of cryptosporidiosis to AIDS patients is very high. A regular monitoring of tap water for investigating cryptosporidium oocytes may be undertaken to screen it.

Isospora belli, a parasite more common in tropical and subtropical regions, is reported to cause chronic or relapsing type of diarrhoea in immunocompromised, associated with weight loss. Demonstration of transparent oocytes in saline mount and fluorescent auramine stain are useful in establishing laboratory diagnosis. As oocytes appear in faeces intermittently, either concentration techniques or multiple specimen collection are mandatory.

Microsporidiosis is recognized as a multisystem disease including gastrointestinal symptoms, renal disease, sinusitis and keratitis in AIDS patients. These are small spore forming obligate, intracellular protozoan parasites that are found in various body sites in wild and domestic animals. Microsporidiosis is an under-diagnosed entity due to the lack of clinical awareness and the difficulty in diagnosis. Infection of the intestinal epithelium with Enterocytozoon bieneusi and Encephalitozoon intestinalis is the most common manifestation of Microsporidiosis in AIDS patients. Clinical manifestations are indistinguishable from that of cryptosporidiosis or isosporiasis in this group of patients. Diarrhoeal stools are watery without blood and not associated with fever. Microsporidia is commonly found in HIV infected patients with CD4 lymphocyte counts less than 50 cells/mm^3 and with complaints of chronic diarrhoea, nausea, malabsorption and severe weight loss.

Microscopy employing various staining techniques like gram stain, Kinyoun's acid fast stain, Giemsa stain, modified trichrome stain as well as fluorescent technique using Calcoflour white were reported to be equally useful for detecting Microsporidia. Genetic probes specific for various species have also been developed. Polymerase Chain Reaction is reported to be superior for speciation of Microsporidia. An ELISA technique for detection of Antibodies to E.cuniculi has been reported.

Toxoplasmosis, a zoonotic disease caused by an intracellular sporozoan parasite, Toxoplasma gondii, has emerged as an important opportunistic infection associated with HIV. However reports from India is scanty. Toxoplasmosis is transmitted by consumption of undercooked meat contaminated with the tissue cyst of *T.gondii.* Another very common mode of transmission is by ingestion of oocyte excreted in the cat's faeces either by close contact with the animal or by playing in the soil contaminated with the oocytes, especially in children.

Toxoplasmosis is a common OI of central nervous system. It infects the brain and can cause raised intracranial pressure, which leads to headache, vomiting and also symptoms like confusion, motor weakness and fever. Disease may progress to seizures, stupor and coma in the absence of treatment. Dissemination, though less common, may affect eyes and may also cause pneumonia.

A study from Thailand, assessed the frequency and distribution of intestinal parasites and the liver fluke Opisthorchis in HIV sero positive and sero negative subjects. A high incidence of liver fluke, Opisthorchis viverrini (19.2%) was found in both the groups followed by **Strongyloides stercoralis** (17.9%) in sero positive group. Many unusual organisms were encountered in this detailed study. In a report, which analyses risk factors for acquiring S.stercoralis infection, corticosteroid therapy and HIV infection were included. Sepsis due to Gram-negative bacilli was noticed as an important complication of Strongyloides hyper infection. Presence of rapidly moving rhabditiform larvae in the faeces can be demonstrated by wet mount examination. High prevalence of Giardiasis and Strongyloidiasis was reported among HIV infected patients in Brazil.

9.4 VIRAL OPPORTUNISTIC INFECTIONS IN HIV INFECTED ADULTS

Viral infections contribute to the significant morbidity and mortality in HIV infected patients. The most common viral opportunistic infection includes Cytomegalovirus (CMV), Varicella zoster virus, Herpes simplex virus (HSV) and Human papilloma virus (HPV). In a study, high prevalence of opportunistic viral infections in late stage HIV disease was noticed. According to the investigators, the incidence was detected in postmortem by histology and cell culture, which correlate with the undiagnosed symptoms in antemortem. In this study viral infections were detected in 72% of all cases comprising CMV as 66%, HSV (11%).

Cytomegalo virus is transmitted through close contact, including sexual contact, body fluids, 'in utero' from infected mother and by breast feeding. Immunocompetent persons may have asymptomatic infection, while in HIV infected persons, like any other infections can have serious clinical manifestation. These symptoms include blind spots in vision, loss of peripheral vision, head ache, mouth ulcers, pain in the abdomen, bloody diarrhea, breathlessness, etc.

Herpes simplex virus cause painful sores on the mouth (fever blisters) genitalia or anal area. Genital herpes is transmitted through sexual contact and the herpes as the mouth is easily spread through kissing. A study stated that recent HSV infection could increase chances of HIV acquisition. **Herpes zoster (shingles)** is caused by varicella virus, the causative agent of chicken pox. It is characterized by the appearance of a unilateral, painful eruption localized on the skin over sensory cells. It occurs as a reactivation of the virus, which remains latent in the sensory root ganglia, following provocation to various stimuli including immunosuppression.

Risk factors for the transmission of **Hepatitis B and C** are similar to that of HIV. Co infection with HIV and hepatitis B and/or C is of great concern in the developed countries. It is reported as upto 89% in some population from developed countries, while 6–33% in India. A study from Manipur reported that 92% of HIV infected intravenous drug users were co-infected with hepatitis C, whereas non-IVDU population the rate varied from 48–21.4%. In the United States, end stage liver disease caused by **HCV** is contributing to the mortality among HIV patients. In a report from South India, prevalence rate of 5% in HBV and 3% in HCV was found in a HIV voluntary counseling and testing center dealing with high-risk individuals.

A study from Kolkata involving female sex workers reported the incidence of **Human papilloma virus (HPV)**, which causes oral, genital and anal warts in 29% of buccal mucosal cells and 63% of cervical cells. **Mollescum contagiosum**, which is characterized by pearly pink papules with central umbilications, is reported to cause disfiguring skin infection in severely immunocompromised individuals. In a histopathologic study of cutaneous lesions in Indian patients with HIV, Molluscum contagiosum was accounted for 14%.

9.5 FUNGAL INFECTIONS IN HIV

Fungal colonization and severe clinical manifestations were reported in HIV infected individuals. **Oral candidiasis** due to Candida albicans was reported as the most frequent fungal infection in HIV infected persons, even upto 20%. Oral lesions aid in the early diagnosis, monitoring of the progress as well as in providing a clue about the immune status of the individual. Oral thrush and the pseudomembranous 'white patches' is associated with more severe immune suppression. Laboratory diagnosis of Candida includes characteristic morphology on Sabourauds Dextrose agar (SDA), gram stain which demonstrates gram positive budding yeast like cells, germ tube formation, chlamydospore formation in corn meal agar.

Pneumocystis jerovecii pneumonia (PCP) is the most common AIDS defining illness in the developed world and its occurrence establishes the diagnosis of AIDS. However in India a very low incidence rate of 0.7–7% was reported. PCP is reported to occur as a co-infection with other pulmonary infection like Tuberculosis, Cryptococcosis and CMV infection in immunosuppressed individuals.

Histoplasma capsulatum, a dimorphic fungi, which grows as yeast form at 37°C and filamentous form at 22°C is also reported as a significant pathogen in dermatopathologic lesions associated with HIV/AIDS patients in the Indian scenario. It is an intracellular infection of reticuloendothelial system, originally reported by Darling in 1905. This fungus is distributed in soil, rotting trees and is abundant in bird faeces. Infection is acquired by inhalation. It resembles tuberculosis and leaves areas of calcification. Disseminated histoplasmosis in HIV is associated with lymphadenopathy, hepatosplenomegaly, fever, anaemia with high mortality. Cutaneous lesions which are granulomatous and ulcerative may be developed on skin and mucosa

Cryptococcus is a pathogenic yeast with a prominent polysaccharide capsule. It is the causative agent of pulmonary cryptococcosis and meningitis. It is a soil saprophyte and present abundantly in the faeces of pigeons and other birds. Infection is usually acquired by inhalation. Cryptococcal meningitis is reported as the most common Central Nervous System manifestation in Indian patients with HIV. It is associated with higher mortality as seen in many studies. Positive blood culture, altered sensorium, high titre of CSF Ag, demonstration of Cryptococcus in CSF by India ink preparation and elevated CSF pressures were considered as poor prognostic factors for cryptococcal meningitis. Cryptococcosis is one of the AIDS defining infections and has been predicted as the 'Mycosis of the future'. Revising the status of HIV co-infection with Cryptococcus, a steep rise was noticed from 20% in 1992–96 period to 49% in 2000–2004 period.

Penicillium marneffei is reported as an important emerging opportunistic pathogen in HIV/AIDS. However the reports from South India are scanty, numerous cases are reported from the eastern states of Indian subcontinent. This fungus is found to be endemic in South East Asia, India and China. It is considered as another AIDS defining illness. This organism was isolated from bamboo rats and from man. Isolates from both the sources shared identical multilocus microsatellite typing. This finding suggests that transmission of *P.marneffei* may occur from rodents to humans or rodents and man are co-infected from common environmental source. Disseminated infection with *P.marneffi* is reported as one of the most common opportunistic infection in HIV infected persons in North Thailand especially in pediatric population. In this study early diagnosis was based on the demonstration of the fungus in skin smear or lymph nodes. The characteristic morphology of colonies on SDA and microscopic appearance will be helpful in the diagnosis. The prompt administration of antifungal therapy employing amphotericin B, fluconazole or ketoconazole improved the outcome. *P.marneffei* is considered as the third most frequent opportunistic pathogen after tuberculosis and cryptococcosis in endemic areas.

9.6 MALIGNANCIES IN HIV/AIDS

People infected with HIV are more vulnerable to certain malignancies like Kaposi's sarcoma, Non Hodgkins Lymphoma, Squamous cell carcinoma and cervical carcinoma and also considered as AIDS defining conditions.

Kaposi's Sarcoma, a malignancy of skin commonly seen in African population is found to occur in homosexual men with HIV/AIDS and the causative agent being human Herpes virus-8. Lesions are found to occur in various parts of the body including skin, lymph nodes with an invasive organ involvement.

Non Hodgkins Lymphoma (NHL) is a cancer of lymph system with different subtypes. The most common subtype with HIV/AIDS are primary central nervous system lymphoma, primary effusion lymphoma or intermediate and high grade lymphoma. Women with HIV/AIDS are at a higher risk of developing Cervical Intraepithelial Neoplasia (CIN), a pre cancerous growth of cells in the cervix associated with Human Papilloma virus (HPV).

Squamous cell carcinoma (SCC) is a malignant tumor of squamous epithelium. It may occur in various parts of the body including skin, lips, mouth, oesophagus, urinary bladder, prostate, lungs, vagina and cervix. This is another form of malignancy reported in HIV/AIDS.

Treatment: Co-trimoxazole is reported as the drug of choice against Toxoplasmosis, Salmonellosis, Pneumocystis carinii, Haemophilus and Staphylococcal infections in patients with HIV. Despite the possibility of

Hyper Sensitivity reactions- rash and fever- to co-trimoxazole, it may be considered as a valuable tool in the management of HIV. It has been recommended by NIH in the treatment guidelines when CD4 counts are less than 200 cells. According to the current WHO guidelines in patients with a CD4 count less than 200 cells/ mm^3, HAART is recommended between 2 weeks and 2 months after initiation of Tuberculosis treatment and it can be deferred if CD4 count is greater than 350 cells. With the widespread use of combination anti retroviral therapy, incidence of refractile diarrhea due to cryptosporidia and microsporidia is remarkably reduced.

In majority of the patients whose CD4 count is below 200 cells, HAART is found to be successful, as it suppresses viral replication and resulting in the reconstitution of the immune system. Administration of Amphotericin B for 2 weeks, which involves renal toxicity, is considered as the treatment for cryptococcal meningitis.

Isoniazid preventive therapy is recommended as the treatment of choice for HIV infected persons at risk of Tuberculosis and those with latent Tuberculosis infection. The DOTS (directly observed short course) treatment strategy recommended by WHO is effective in HIV infected persons. Immune reconstitution syndrome (IRS), a paradoxical worsening of clinical status after initiation of HAART in patients with active opportunistic infections like Tuberculosis also should be taken into consideration by the physician.

9.7 CONCLUSION

HIV has got diverse clinical manifestations with multiple organ involvement. Due to the remarkable immunosuppression, OIs of various aetiology is expected to occur in these patients and the profile vary considerably from person to person and from region to region. As it is the greatest challenge to public health, high level of alertness is required in the early detection of HIV, chemoprophylaxis for opportunistic infections and the availability of HAART. This in turn will definitely reduce the morbidity and mortality in these patients. Special laboratory facility exclusively for the infections in immunocompromised patients should be the dream project of each Microbiologist as well as the clinician.

KEYWORDS

- **AIDS**
- **CD4 Count**
- **HAART**
- **human immunodeficiency virus (HIV)**
- **lymphadenopathy**
- **opportunistic Infections**

REFERENCES

Ayyagari, A., Sharma, A. K., Prasad, K. N., Spectrum of opportunistic infections in HIV infected cases in a tertiary care hospital. *Indian J Med Microbiol* 1999, 17, 78–80.

Banerjee, U., Dutta, K., Majumdar, T., Gupta K. Cryptococcosis in India: Awakening of a giant? *Med Mycol* 2001; *39*, 51–67.

Banergee U. Progress in diagnosis of opportunistic infections in HIV/AIDS. *Indian J Med Res* 2005, 121, 395–406.

Bhattacharya, S., Badrinath, S., Hamide, A., Sujatha S. Coinfection with hepatitis C virus and human immunodeficiency virus among patients with sexually transmitted diseases in Pondicherry, South India. *Indian J Pathol Microbiol* 2003, *46,* 495–7.

Deshmukh, S. D., Ghaisas, M. V., Rane, S. R., Bapat, V. M., *Pneumocystis carinii* pneumonia and its association with other opportunistic infections in AIDS – an autopsy report of five cases. *Indian J Pathol Microbiol* 2003, *46,* 207–211.

Goodgame, R. W., Understanding intestinal spore forming protozoa: *Cryptosporidia, Microsporidia, Isospora* and *Cyclospora*. *Ann Intern Med* 1996: *124,* 429–41.

Hammouda, N. A., Sadaka, H. A., El-Gebaly, W. M., El-Nassery, S. M., Opportunistic intestinal protozoa in chronic diarrhoeic immunosuppressed patients. *J Egypt Soc Parasitol* 1996; *26,* 143–53.

Jain, S. K., Aggarwal, J. K., Rajpal, I., Baveja U. Prevalence of HIV infection amongst tuberculosis patients in Delhi-A sentinel surveillance study. *Indian J Tuber* 2000; *47:* 21–6.

Janoff, E. N., Smith, P. D., Prospective on gastrointestinal infections in AIDS. *Gastroenterol Clin North Am* 1988; *17:* 451–63.

Jones, J. L., Hanson, D. L., Dworkin, M. S., Alderton, D. L., Fleming, P. L., Kaplan, J. E., Ward J. Surveillance for AIDS defining opportunistic illnesses, 1992–1997. *MMWR CDC Surveill Summ* 1999; *48,* 1–22.

Khanna, N., Chandramuki, A., Desai, A., Ravi V. Cryptococcal infections of the central nervous system: an analysis of predisposing factors, laboratory findings and outcome in patients from South India with special reference to HIV infection. *J Med Microbiol* 1996; *45,* 376–9.

Kumaraswamy, N., Vallabhaneni, S., Flanigan, T. P., Mayer, K. H., Solomon S. Clinical profile of HIV in India. *Indian J Med Res* 2005; 25, 377–94

Lanjewar, D. N., Bhosale, A., Iyer A. Spectrum of dermatopathologic lesions associated with HIV/AIDS in India. *Indian J Pathol Microbiol* 2002;*45*:293–8.

Morbidity and Mortality Weekly Report. 1993 Revised Classification System for HIV Infection and Expanded Surveillance Case Definition for AIDS Among Adolescents and Adults. *MMWR* 1993; 41 RR-17: http://www.cdc.gov/ view/mmwrhtml/00018871.htm. accessed on April 20th, 2013.

Mamidi, A., DeSimone, J. Q., Pomerantz, R. J., Central nervous system infections in individuals with HIV-1 infection. *J Neurovirol* 2002; *8*, 158–67.

Meisheri, Y. V., Mehta, S., Patel U. A prospective study of seroprevalence of toxoplasmosis in general population, and in HIV/AIDS patients in Bombay, India. *J Postgrad Med* 1997; *43,* 93–7.

Misra, S. N., Sengupta, D., Satpathy, S. K., AIDS in India: Recent trends in opportunistic infections. *Southeast Asian J TropMed Public Health* 1998; *29*, 373–6.

Mukhopadhya, A., Ramakrishna, B. S., Kang, G., Pulimood, A. B., Mathan, MM, Zacharian, A., *et al.* Enteric pathogens in southern Indian HIV-infected patients with & without diarrhea. *Indian J Med Res* 1999; *109*, 85–9.

Ranganathan, K., Reddy, B. V. R., Kumarasamy, N., Solomon, S., Viswanathan, R., Johnson, N. W. Oral lesions and conditions associated with human immunodeficiency virus infection in 300 South Indian patients. *Oral Dis* 2000, *6*, 152–157.

Reynolds, S. J., Risbud, A. R., Shepherd, M. E. Recent herpes simplex virus type 2 infection and the risk of human immunodeficiency virus type 1 acquisition in India. *J Infect Dis* 2003; *187,* 1513–21.

Shelburne, S. A., Hamill, R. J., Rodriguez-Barradas, M. C., Greenberg, S. B., Atmar, R. L., Musher, D. W., et al. Immune reconstitution inflammatory syndrome: emergence of a unique syndrome during highly active antiretroviral therapy. *Medicine* 2002, *81*, 213–227.

Singh, A., Bairy, I., Shivananda, P. G. Spectrum of opportunistic infections in AIDS cases. Indian J Med Sci 2003, 57, 16–21.

Thomas, Q. C. HIV-AIDS Related Problems in Developing Countries. In: Doland Amstrong Cohen eds: Infectious Diseases, London: Mosby; 1999.

Vanittanakom, N., Coopes, C. R., Fisher, M. C., Sirisanthana, T. *P. marneffei* infection and recent advances in the epidemiology and molecular biology aspects. Clin Microbiol Rev 2006, 19, 95–110.

Weber, R., Bryan, R. T., Schwartz, D. A., et al., Human microsporidial infections. Clin Microbiol Rev 1994, 7, 426–461.

CHAPTER 10

RELATIONSHIP BETWEEN PSYCHO-SPIRITUAL WELL-BEING AND PHYSICAL-ILLNESS BEHAVIOR IN HIV SEROPOSITIVE INDIVIDUALS

MWIYA LIAMUNGA IMASIKU

CONTENTS

ABSTRACT

Objective: The objective of the present study was to compare spiritual wellbeing and physical-illness behavior in HIV seropositive individuals and individuals from the normative population who are not HIV positive to find out if there is any relationship. A growing body of evidence demonstrates a significant relationship between spirituality and health. HIV-infected individuals often find new meaning and purpose for their lives while establishing new connections and strengthening old ones (Philips, 2006).

Design: This multiple group design was used to assess the intensity and impact of spiritual well-being in 120 HIV seropositive individuals and individuals from the normative population. The study utilized a cohort method, which lasted a period of four years. Spiritual well-being was found to be a significant factor related to physical health status.

Main outcome measure: Spiritual wellbeing and physical wellbeing.

Results: The descriptive analysis of existential wellbeing among Zambian and Indian participants shows that 70% of HIV seropositive individuals from Zambia and India who did not have a sense of relatedness to a transcendent dimension or something greater than the known and permanent self, [religious wellbeing], had more experiences of illness behavior [somatic symptoms]. It was also found that 66% of HIV seropositive individuals who did not have a sense of life purpose and life satisfaction had more experiences of illness behavior [somatic symptoms] than their counterparts who found that their lives were meaningful and purposeful.

The 'F' ratio for diagnosis was significant at 0.05 while the F ratio for nationality was significant at 0.01 levels indicating that diagnosis and nationality independently influenced spiritual well-being but not interactively. Correlation between the level of physical illness behavior [somatic symptoms] and religious wellbeing was found to be -0.103 while that for existential well-being was –0.419. This indicated that physical illness behavior is negatively correlated with both existential and religious well-being. However, the correlation between existential well-being and physical illness behavior was higher than the correlation between religious wellbeing and physical illness behavior.

Conclusion: When the level of spiritual well-being is high by an individual, then it's likely that physical illness behavior will be low because those who had adequate spiritual well-being had less illness behavior. The results therefore illustrate that less levels of spiritual well-being and physical illness behavior seem to co-exist. The findings suggest that spiritual well-being and physical health quality need to be assessed so appropriate interventions can be implemented to improve health outcomes in this population. Therefore provision of adequate spiritual well-being should be included in the management and treatment of HIV seropositive individuals in order to enhance their quality of life.

10.1 INTRODUCTION

Increasingly, health status is important in HIV-infected individuals due to the chronic nature of the disease, and it is a critical factor to consider in the introduction and implementation of interventions. It is probably important to note that for many lay individuals in Zambia and elsewhere, a healthy person is probably perceived to be the individual who is fat and who does not show any signs of physical ailments. However, the truth is that health status is a multi-factorial construct consisting of not just physical and psychological components but also spiritual wellbeing. It is therefore necessary that health PR actioners of different categories need to be cognizant of the fact that they are commissioned to treat people as whole persons- body, mind, and spirit. In the west, there is an observation that biochemical medical medicine fails to treat patients as whole living persons because it does not account for the spiritual dimensions of patients (Phillips, 2006). That is why the Mind-body medicine does not just focus on the interactions among the brain, body, and behavior but also on how spiritual factors can directly affect health (Akiiki and Ngare, 2006). No wonder the WHO does not define health as just physical and psychological well-being but also as spiritual well-being (World Health Organization, 1976). Many scholars mighty therefore wonder about the implications of the absence of spiritual well-being to the physical health many people living with the HIV during the course of their illness. It is a well known fact that Anti Retroviral drugs [ARVs] can prolong life.

> *"But what's the use of just prolonging life or leading a miserable life for a long time without any value, purpose or meaning attached to it? Is it not more ideal to prolong life using ARVs and also at the same time add life to years through improving their spiritual well-being?" (Imasiku, 2009)*

The significance of spiritual well-being in the management of HIV and AIDS is because maladaptive beliefs HIV seropositive individuals have can make treatment complex. These beliefs have not been empirically verified, however, health workers need to be aware of them because they influence people's perceptions of the causes of illness and the treatments sought (Phillips and Mock, 2006).

Since the HIV pandemic poses a threat to national development as it decimates the productive population, leaving only vulnerable and poorly looked after future generations (http://www.unicef.org/sitan/files/UNICEF_GRZ_Situation_Analysis_2008.pdf) the purpose of this article is to analyze the impact of spiritual well-being on HIV seropositive individuals in order to manage HIV and AIDS from all fronts. This is important because even if it appears that there is a marked decline in the rate of new AIDS cases because of the life-prolonging benefits of antiretroviral therapy, HIV disease challenges every aspect of an HIV-infected individual's life, often contributing to spiritual distress. One study revealed that, over time, HIV seropositive individuals often find new meaning and purpose for their lives and strengthen old connections with their spiritual belief systems (McCormick et al., 2001). Other studies indicated that Spirituality is significantly associated with beneficial outcomes in HIV-infected individuals (Woods et al., 1999).

The term spirituality is synonymous with the word spiritual wellbeing. Spiritual well-being is divided into two components: religious well-being and existential well-being. While the term "religious wellbeing" is a specific term used to refers to a sense of relatedness to a transcendent dimension or with a higher being which is greater than the known and permanent self, existential well-being is another specific component of spiritual wellbeing which refers to a sense of life purpose and life satisfaction concerned with a person's ethical and moral intentions and behaviors that affect the physical world and human interactions (Paloutzian and Ellison, 1982).

In a study of 52 men living with HIV, it was revealed that spiritual well-being is associated with health and resilience in HIV-infected individuals. Those who had high spiritual well-being had effective coping strategies (Tuck et al., 2001). It therefore means that there might be a negative relationship between existential well-being and emotional-focused coping which is positive and adaptive (Tuck et al., 2001). While one study suggested that resilience was significantly related to religious affiliation, religious faith, number of symptoms, level of physical functioning, and socioeconomic status (Flannelly and Inouye, 2001), another study indicates that of 96 HIV infected individuals, resilience was found to be significantly related to not just one component called religious well-being but spirituality generally. It was further revealed that a combination of religious wellbeing and existential wellbeing and social support explained the variance in resilience (Poblete, 2000).

Another study by Kenneth Phillips and Kathryn Mock (2006) revealed that Spiritual well-being was found to be a significant factor related to sleep quality, mental health and physical health status. In their study, it was reported that every participant had sleep disturbance. Other studies by Darko et al. (1998), Gardner et al. (1997), Mock et al. (2002) and Nokes et al. (2001) revealed that sleep disturbance is a frequent and troublesome complication of HIV infection with prevalence rates reaching over 70%. Poor sleep quality has been associated with increased morbidity in HIV-infected individuals. These findings suggest that spiritual well-being and sleep quality need to be assessed so that appropriate interventions can be implemented to improve health outcomes in this population.

In a study of 80 HIV-infected females, it was found that spiritual perspective, defined as "one's frame of reference consisting of beliefs and attitudes of connectedness with something or someone greater than self", was positively correlated with mastery over stress, which is defined as successfully coping with stress and experiencing change, growth, acceptance, and enhanced life meaning (Gray and Cason, 2002). That is why the concept of spiritual well-being is also associated with reduced feelings of death distress in patients with life-threatening illness (Chibnall et al., 2002). Although association does not show causality, these results suggest that health, resilience, and spiritual well-being are closely related.

Why is there focus on spiritual wellbeing in the treatment of HIV seropositive individuals? The focus goes back to the principles on which WHO was founded in that health is not simply the absence of physical illness, because both religious faith and religious affiliation contributed to improved health in HIV-infected individuals (Chibnall et al., 2002). Research thus reveals that what may contribute to health status by providing meaning and purpose in life is existential well-being (Flannelly and Inouye, 2001). The present study is worth doing to find out if these research outcomes from the west and elsewhere also apply to developing countries like Zambia and India.

The current study was conducted to compare the spiritual wellbeing of HIV seropositive individuals and individuals from the normative population who are not HIV positive to find out how the components of spiritual wellbeing, namely, religious well-being and existential wellbeing affects physical well-being in India and Zambia.

10.2 HYPOTHESES

The foregoing theoretical analysis led to the formation of the following hypothesis.

- Spiritual wellbeing will influence physical health of HIV seropositive individuals.
- Physical illness behavior is positively correlated to inadequate spiritual wellbeing.
- Diagnosis and nationality will independently and interactively influence the spiritual dimensions of well-being.

10.3 METHODOLOGY

10.3.1 RESEARCH DESIGN

A descriptive multiple group design which is correlational in nature was used to test the relationships between spiritual well-being and physical wellbeing in a sample of HIV seropositive individuals and individuals from the normative population. Subsequently, the relationships between religious well-being and existential well-being and somatic symptoms were tested.

10.3.2 SAMPLE AND SETTING

These analyses are from a study of psychoneuroimmunological correlates of religious well-being and existential well-being and somatic symptoms in HIV disease. The participants were 120 seropositive individuals between 18 and 50 years of age. In Zambia participants (n=60) were recruited from University Teaching Hospital and Kara counseling center while in India they came from Snehadhan and Karnataka Network for people Living with HIV [KNP+].

HIV seropositive individuals and those from the normative population were selected from the hospital setting or other health institutions to establish diagnosis of their HIV status. Individuals from the normative population were basically those individuals who had gone there for the purpose of voluntary counseling and testing in order to know their status. The criteria used to select these institutions depended on the availability of the required respondents in those institutions.

The sampling technique, which was used in the research was stratified sampling because the population was categorized on the basis of nationality (Zambians and Indians) and on the basis of diagnosis (individuals from the normative population and HIV seropositive individuals). Randomized sampling was not used because of the nature of privacy in matters relating to HIV especially in India. Therefore only those individuals who met the inclusion and exclusion criteria and had concerted to be available for the present study were selected. Exclusion criteria for HIV seropositive individuals selected included: (i) being too ill to come to the assessment center; (ii) being a current intravenous drug user; (iii) inability to speak English; (iv) previous history of psychiatric or neurological consultations and mental deficiencies; (v) unwilliness to be part of the study; (vi) exposure to a similar study before.

All the participants were males between the ages of 18 and 50 years in order to come up with a more homogeneous population and cut down on the effects on extravenous variables such as sex. The ages between 18 and 50 was selected because the HIV prevalence in both India and Zambia is mostly within this range. The sample was ethnically diverse in the sense that it included two broad and different ethnicities; Indians and Zambians because with a more culturally diverse HIV+ population, we could better understand what factors specifically contribute to this important construct. Homogenity in terms of sex and literacy levels was enhanced by ensuring that HIV seropositive individuals selected for the present study were all diagnosed in the first one year and their educational background/status was at least grade ten.

This chapter involved administering the spiritual wellbeing questionnaire to determine levels of religious and existential wellbeing and somatic symptoms of the subjects.

10.3.3. INSTRUMENTS

I. **Consent form:** A consent form was developed for the present study. This is a written consent form which elicited information that the respondents have agreed to be part of the study. Prior to the administration of all parameters, the willingness of the subjects was ascertained and they were made to sign a consent form.
II. **Information schedule:** This information schedule, which was semi-structured, was specially developed to collect data relevant for the study and ascertain certain socio-demographic details.
III. **Markers of HIV Illness Stage:** Health status was measured by the Symptoms questionnaire (SQ). This instrument was developed from the Medical Outcomes Study by Kellner in 1986 with the aim of making the scales more sensitive for clinical research. The SQ consists of 92 items of which 68 items indicate symptoms (subscales) and 24 items are antonyms of some of the symptoms that indicate physical well-being subscales, which form the basis for Somatic Symptoms. Physical illness behavior implies somatic symptoms, which in the present study are indicators of physical illnesses such as head pains, cramps, muscle pains, upset bowels, nausea, weak arms and legs. The SQ assesses components of health status including: physical functioning, role limitations due to physical health problems, bodily pain (Paloutzian and Ellison, 1982).
IV. **Spiritual Well-Being:** The Spiritual Well-being Scale (SWBS) was used to measure spiritual well-being (Paloutzian and Ellison, 1982). The SWBS is a 20-item self-report instrument with two subscales, the Religious Wellbeing Scale (RWBS) and the Existential Well-being Scale (EWBS).

The RWBS contains ten items that assess an individual's relationship with God, and there are ten items that assess a person's relationship with the physical world and other individuals (EWBS). Each item is rated on a six-point scale that ranges from (1) strongly disagree to (6) strongly agree. Regarding reliability of the SWBS alpha coefficients greater than 0.82 for the total instrument support the reliability of this instrument (Paloutzian and Ellison, 1982). The item content of this instrument supports its face validity, and construct validity has been supported by factor analysis (Paloutzian and Ellison, 1982). A higher score indicates greater spiritual well-being.

10.3.4 PROCEDURE AND ETHICAL CONSIDERATION

Before the study commenced, the Ethics Committee approval was obtained. Therefore, all necessary ethical guidelines were considered in this research. Subject interviews and answering of questionnaires took place at the institutions were they were recruited [University Teaching Hospital and Kara Hope house in Zambia and Network for people living with HIV in Karnataka, India]. Full explanations about the purpose of the study were made to participants and informed consent was obtained from those who agreed to participate in the study. Among other things, the participants were told everything regarding what would be required of them if they agreed to participate. The necessary assurance about the voluntary nature of the study that they would suffer no loss of benefits if they chose not to participate was given. Confidentiality was assured. After the informed consent was read to the participants, they were asked to sign the informed consent statement. An investigator or a trained interviewer conducted the interviews and recorded the participants' answers verbatim.

10.3.5 DATA ANALYSIS

The data collected during the main study was subjected to statistical analysis to compare HIV seropositive individuals, and individuals from the normative population and verify the hypotheses stated. The protocols were all hand-scored. The raw scores were determined for the well-being dimensions based on the manuals and then they were subjected to various statistical analyses to arrive at meaningful interpretation. The hypotheses were tested by analyzing the inter-group and intra group data.

The 1st hypothesis stated that physical Illness behavior is negatively correlated to spiritual well-being. In order to carry out the investigation, the symptoms questionnaire (Kellner, 1986) was administered and the Pearson coefficient of correlations were computed.

The 2nd hypothesis states that diagnosis and nationality will independently and interactively influence the spiritual dimensions of well-being, i.e., HIV seropositive individuals, and individuals from the normative population in India and Zambia will differ on spiritual well-being. To carry out this investigation, the spiritual well-being scale by Paloutzian and Ellison (1982) was administered and the results were subjected to a 2-way ANOVA(Paloutzian and Ellison, 1982).

The data were analyzed in SPSS 11.5 (Chicago, IL, United States of America). Chi-square analysis was used to compare the three groups to ensure that they were homogenous. When appropriate, Pearson correlations to assess the relationship among diagnoses, and physical illness behavior.

Frequencies and percentages were calculated for each of the categorical variables; Means and standard deviations were calculated for each of the continuous variables. The research questions were tested using Pearson's coefficient of correlation (Pearson's *r*). All questions were answered on a voluntary basis. An alpha level of 0.05 was established as the level of statistical significance.

10.4 RESULTS

10.4.1 STUDY SAMPLE

The study sample consisted of 120 HIV seropositive individuals, and normal individuals from Zambian and Indian institutions who were classified into groups based on their diagnosis. Table 10.1 provides a summary of the sample distribution and characteristics of participants. The majority of the participants were between the ages of 14 and 50. The reason for selecting this age group is because it constitutes both the HIV prevalence [the percentage of persons ages 15 to 49 who are HIV infected] and HIV incidence [the percentage of uninfected 15 to 49 year olds who become newly infected each year].

TABLE 10.1 The Sample Distribution of the Study

	HIV seropositive Individuals	Age		Institution	Individuals from the normative population	Total
Zambia	30	Age Range	Frequency	UTH	**30**	**60**
		20–29	43.3	KCC		
		30–39	36.6	NZP+		
		40–50	20			
		Mean	33.1			
		S.D.	7.1			

TABLE 10.1 *(Continued)*

	HIV seropositive Individuals	Age		Institution	Individuals from the normative population	Total
India	30	Age Range	Fre-quency		**30**	**60**
		20–29	43.3	Snehadhan		
		30–39	36.6	KNP+		
		40–50	20			
		Mean	34.9			
		S.D.	6.19			
	60				60	120

#KNP+: Karnataka network for people living with HIV.

UTH: University Teaching Hospital.

NZP+: Network of Zambian people with HIV/AIDS.

KCC: KARA Counseling Centre (Hope House).

To check whether the samples in terms of annual family income are comparable, chi square analysis was done and the Table 10.2 gives the results.

TABLE 10.2 Family Income of Respondents from Zambia and India

Family annual income	India			Zambia		
	HIV(+)	Normal		HIV(+)	Normal	
Below 219 US Dollars	7	5	24	9	16	32
219 US Dollars to 879 US Dollars	13	14	40	13	7	33
879 US Dollars to 1,319 US Dollars	10	11	26	8	7	25
Total	30	30	90	30	30	90
	X^2=3.200 Not significant DF=2			X^2=1.867 Not significant DF=2		

Since all the above X^2 values are non-significant, we can therefore conclude that the groups are homogenous in terms of the significant demographic variables of education and annual family income. The rationale for selecting a homogeneous group in terms of social-economic status, [respondents who are above poverty line] is because, poverty has been reported to play a significant role in not only the spread of HIV but also in expediting the progression of HIV into AIDS (http://www.unicef.org/sitan/files/UNICEF_GRZ_Situation_Analysis_2008.pdf).

10.4.2 SPIRITUAL WELLBEING OF THE ZAMBIAN AND INDIAN SAMPLE

TABLE 10.3 Spiritual Well-Being in HIV+, Normal Individuals in India and Zambia

Variable	**Diagnosis Nationality**	**HIV+**		**Normal**		**Total**	
		Mean	SD	Mean	SD	Mean	SD
Religious	Zambia	34.70	5.66	38.67	7.99	35.71	6.49
Well-being	India	37.93	10.58	42.60	12.73	40.09	10.9
Total		36.32	8.57	40.63	10.72	37.90	9.26
				Df	F	Sig	
	Main Effects	Diagnosis		2	4.288	.015*	
		Nationality		1	10.91	.001**	
	2-way Interaction	Nationality x Diagnosis		2	0.383	.683 NS	
		HIV+		Normal		Total	
		Mean	SD	Mean	SD	Mean	SD
Existential	Zambia	36.93	9.14	43.30	10.26	38.22	10.1
Well-being	India	35.37	6.24	40.57	7.74	37.01	6.44
Total		36.15	7.80	41.93	9.12	37.62	8.44
				Df	F	Sig	
	Main Effects	Diagnosis		2	13.884	.000*	
		Nationality		1	1.057	.305	
	2-way Interaction	Nationality x Diagnosis		2	.717	.490 NS	

** P<.01; * P<.05, NS, Not significant.

10.4.3 RELIGIOUS WELL-BEING

This term refers to a sense of relatedness to a transcendent dimension or something greater than the known and permanent self.

Table 10.3 shows that on existential wellbeing, HIV seropositive individuals from Zambia had a mean of 36.93±9.14 while HIV seropositive individuals from India had a mean of 35.37 ± 6.24. Regarding religious well-being, HIV seropositive individuals from Zambia had a mean of 34.70±5.66 while HIV seropositive individuals from India had a mean of 37.93±10.58, respectively. There were no significant differences in spirituality between HIV seropositive individuals from Zambia compared to HIV seropositive individuals from India. However, it is interesting to note that 70% of HIV seropositive individuals from Zambia and India who did not have a sense of relatedness to a transcendent dimension or something greater than the known and permanent self [God] had more experiences of illness behavior [somatic symptoms]. It was also found that 66% of HIV seropositive individuals who did not have a sense of life purpose and life satisfaction had more experiences of illness behavior [somatic symptoms] than their counterparts who found that their lives were meaningful and purposeful.

The 'F' ratio for the main effect- diagnosis is significant at 0.05 and the F ratio for nationality is significant at 0.01 level. The F ratio for interaction effects between nationality and diagnosis for religious well-being is not however significant. Looking at the Table 10.4, we see that it indicates that RWB differs across diagnosis. Normal have higher RWB. This indicates that diagnosis and nationality independently influence spiritual well-being but not interactively. So, the hypothesis is partially accepted.

10.4.4 EXISTENTIAL WELL-BEING

EWB refers to that part of the individual which reaches out and strives for meaning and purpose in life. The 'F' ratio for the main effects-diagnosis is significant at 0.05 and the F ratio for nationality is not significant. The F ratio for interaction effects is also found to be significant. Looking at the Table 10.4, we see that it indicates that EWB differs across diagnosis. Although there are variations according to diagnosis, existential wellbeing among HIV seropositive individuals makes a big difference. This indicates that diagnosis and nationality independently influence spiritual well-being but not interactively. So, the hypothesis is partially accepted.

TABLE 10.5 Correlation Between the Level of Physical Illness Behavior [Somatic Symptoms] and Spiritual Well-Being

Factors correlated to Somatic symptoms	Correlation Coefficient
Religious well-being	–0.103
Existential well-being	–0.419**

Religious well-being and existential well-being: The correlation between physical illness behavior and the existential well-being components was found to be negatively but significantly correlated. However, although the correlation between religious well-being and physical illness behavior was also found to be negative, it was not significant (–0.1.3).

TABLE 10.6 "t" Scores Comparing the Different Subgroups on Religious Well-Being and Existential Well-Being

Scale	Religious well-being			Existential Well-being		
Diagnosis	Zambia	India	Total	Zambia	India	Total
HIV & Normals individuals	–1.54NS	–2.22*	–2.44*	–2.54**	–2.86**	3.73**

The 'F' ratio for the main effect- diagnosis is significant at 0.05 and the F ratio for nationality is significant at 0.01 level. The F ratio for interaction effects between nationality and diagnosis for religious well-being is not however significant. Looking at the Table 10.6, we see that it indicates that RWB differs across diagnosis. Normals have higher RWB. Illness makes a difference. When Pearson correlation was done, it was found that existential well-being (r =0.89, p <0.0001) and religious well-being (r =0.80, p <0.0001) were found to be significantly correlated with each other (r=0.66, p <0.0001). Existential wellbeing was significantly related to illness behavior (r =-0.39, p <0.0031) and to total spiritual well-being (r =–0.19, p =0.031), but not to religious well-being (p=.291, r =–0.10).

10.5 DISCUSSION

It was observed that although the variations of spiritual wellbeing components on the basis of nationality were not very significant, this finding indicated that those respondents from either Zambia or India who obtained high spiritual wellbeing had less physical illness behavioral problems in the form of somatic symptoms. The result that most of HIV seropositive individuals from Zambia and India who did not have a sense of relatedness to a transcendent dimension or something greater than the known and permanent self [God] had more experiences of illness behavior [somatic symptoms] is in line with the findings of other studies that spiritual well-being is associated with health and resilience in HIV-infected individuals and that those who had high spiritual well-being had effective coping strategies (Paloutzian and Ellison, 1982). Another implication of this result is that compared to religious wellbeing, the correlation between existential wellbeing and physical illness behavior was stronger. This result shows that although both constructs are important components of spiritual well-being, they are distinct. This therefore suggest that existential well-being or one's estimate of the meaningfulness of one's life is one of the most important "resistance assert" which probably plays an important role in reducing or preventing somatic symptoms. If this argument were anything scientific to go by, then measuring spirituality among other factors in individuals suffering from chronic illness would be an important addition to the management and treatment of HIV seropositive individuals in order to enhance their quality of life. Since the term "quality of life" is a general term used to describe all aspects of an individual's well-being, there is need for serious inclusion of the dimension of spiritual wellbeing in the management of HIV and AIDS as regards research, assessment and interventions.

Since as in other studies cited earlier, the present study indicates stronger correlation between existential wellbeing and physical illness behavior than with religious wellbeing, the hypothesis that Spiritual wellbeing will influence physical health of HIV seropositive individuals is partially accepted. This observation is the main contribution of the present study because it contradict the findings of those studies which stated that there is always a significant relationship between RWB and physical illness behavior. This could be so basically because the meaning and purpose of life depends on the quality of relationship an individual has with God and not merely believing in God (Davison et al., 2010). It can thus be concluded that the lower levels of physical illness behavior in some HIV-Infected individuals in the present study was due to the higher level of spirituality in general among the respondents as was observed in another study among African Americans (Paloutzian and Ellison, 1982).

10.5.1 LIMITATION OF THE STUDY

The limitation of religious wellbeing is undoubtedly found in the variations of the doctrines of different inter and intra denominations. Diagnosis and nationality will independently and interactively influence the spiritual dimensions of well-being.

10.5.2 RECOMMENDATION OF THE STUDY

The findings of the present study therefore supports the clinical implication as was reported by Kenneth Phillips (2006) that due to the physically damaging nature of HIV disease, the facet of spiritual well-being also needs to be assessed during patient care visits. Clinicians can do this by assessing not only stress levels, and coping mechanisms but also concerns about the meaning and purpose of life, energy levels, and role fulfillment. This suggestion is emphasized because additional assessments and interventions are important in holistically treating the HIV patient.

The findings of the present study also supports the clinical implication as was reported by Kathryn Mock (2006) that as with all patients afflicted with chronic and potentially fatal diseases, nurses need to actively promote improved health status through a variety of interventions. Although the hope for a cure or the halting of progression of the disease is often foremost in the patient's mind, nurses must consider all factors that can affect the individual's daily life and prognosis. Spirituality has long been recognized as a resource that can improve physiologic and psychological factors for patients with chronic illness. Clinicians often are uncomfortable discussing spirituality with their patients, yet the assessment may begin by simply asking the patient if they attend religious services. Patients may need to be reminded to resume their traditional religious activities or to seek out new connections with established spiritual resources. Religious services can be a resource for individuals in many ways by providing social support, structure and stability, meaning and purpose in life, and an outlet for spiritual expression.

Individuals also may need to be supported in their existential awareness and concerns in order to maximize their health. This may simply mean encouraging patients to start or continue volunteer work that they have done in the past. Participating in research studies, helping out at a service organization, or contributing to a food drive are all activities that support having a purpose in life and self-transcendence, and could improve existential spiritual well-being. Creative activities can improve an individual's existential well-being: The nurse might suggest keeping a journal, writing a poem, drawing, or gardening, as ways of tapping into creative energies. Creative activities such as these may even help a person find purpose and meaning in life and help with an individual's need to create a legacy. In conclusion, a wide range of physical, mental, and psychosocial factors influence quality of life in HIV-infected individuals. Additional studies are needed to elucidate the relationship between spirituality and health in this and other clinical populations.

10.6 CONCLUSION

Due to the physically damaging nature of HIV disease, many facets of an individual's health need to be assessed during patient care visits such as concerns about the meaning and purpose of life. Therefore interventions targeting enhancement of existential wellbeing are important in holistically treating the HIV patient. Though emphasis is on existential wellbeing, it is important to note that in most cases an individual may not afford to have sound existential wellbeing without first having good religious wellbeing. This study therefore joins the rest of the studies in attesting to the fact that spirituality is indeed a resource that can improve physiologic factors for patients with chronic illness.

KEYWORDS

- **HIV seropositive**
- **somatic symptoms**
- **spiritual well-being**

REFERENCES

Akiiki, Ngare, D., et al. (2006) The African Textbook of Clinical and Mental Health Nairobi.

Chibnall, J. T., Videen, S. D., Duckro, P. N., Miller, D. K. (2002). Psychosocial spiritual correlates of death distress in patients with life-threatening medical conditions.

Flannelly, L. T., Inouye, J. (2001). Relationships of religion, health status, and socioeconomic status to the quality of life of individuals who are HIV positive. *Issues in Mental Health Nursing, 22*, 253–272.

Gray, J., Cason, C. L. (2002). Mastery over stress among women with HIV/AIDS. *Journal of the Association of Nurses in AIDS Care, 13*, 43–51.

http://www.unicef.org/sitan/files/UNICEF_GRZ_Situation_Analysis_2008.pdf.

Imasiku, M. L. (2009). Social Wellbeing Predictor of illness Behavior among HIV seropositive individuals. Medical Journal of Zambia 36 (4), 157–164.

Kenneth, D. Phillips Kathryn S. Mock (2006). Spiritual Well-being, Sleep disturbance, and Mental and Physical Health status in HIV-infected individuals. South *Carolina.*

McCormick, D. P., Holder, B., Wetsel, M. A., Cawthon, T.W. (2001). Spirituality and HIV disease: An integrated perspective. *Journal of the Association of Nurses in AIDS Care, 12*, 58–65.

Paloutzian, R. F., Ellison, C. W. (1982). Loneliness, spiritual well-being, and quality of life. In L. A. Peplau, D. Perlman (Eds.), *Loneliness: A sourcebook of current theory, research and therapy* (pp. 224–237). New York: Wiley.

Philips (2006). *Issues in Mental Health Nursing*, 27, 125–139, Taylor & Francis Group.

Poblete, S. A. (2000). *Relationship of spirituality, social support, reciprocity and conflict to resilience in individuals diagnosed with HIV*. Newark, NJ: Rutgers, The State University of New Jersey. *Spiritual Well-Being in HIV-Infected Individuals* 139.

Sara N. Davison et al. (2010), Existential and Religious Dimensions of Spirituality and Their Relationship with Health-Related Quality of Life in Chronic Kidney Disease *CJASN ePress.*

Tuck, I., McCain, N. L., Elswick, R. K., Jr. (2001). Spirituality and psychosocial factors in persons living with HIV. *Journal of Advanced Nursing, 33*, 776–783.

Woods, T. E., Antoni, M. H., Ironson, G. H., Kling, D. W. (1999). Religiosity is associated with affective and immune status in symptomatic HIV-infected gay men. *Journal of Psychosomatic Research, 46*, 165–176.

World Health Organization (1976). Health aspects of Human Rights, WHO, Geneva.

CHAPTER 11

ADVANCES IN THE UNDERSTANDING AND TREATMENT OF CHILDREN WITH CANCER AND HIV INFECTION

DANIELA CRISTINA STEFAN

CONTENTS

ABSTRACT

More than 95% on the new infections and HIV/AIDS-associated deaths occur in developing countries.

Children account for 14% of all new infections and 18% of all deaths due to the HIV. Children infected with the retrovirus have a much higher probability to develop a malignancy during the childhood.

The most common HIV defining and related cancers in children are: Kaposi sarcoma (KS), Burkitt lymphoma (BL) and leyomyosarcoma (mainly in developed countries).

Despite progress in the treatment and survival of childhood cancer globally, the need for adapted protocols and randomized trials for HIV related malignancies in children remain a priority.

11.1 CHILDREN WITH HIV

Recent World Health Organization data showed an estimated 34 million people worldwide living with HIV infection at the end of 2011 with a concerning 2.5 million people being newly infected by the end of that same year (HIV/AIDS Fact sheet no. 360, 2012). In the pediatric population there were 3.4 million children below the age of 15 infected globally, with 390,000 new infections which occurred in 2010 alone (Global HIV/AIDS Response: Update 2011). Data however suggest that HIV infection among children is decreasing, due to improved access to preventive medication perinatally. About one fourth as many new cases of HIV infection among children are believed to have occurred in 2010 as in 2005 (Global HIV/AIDS Response: Update 2011).

AIDS related deaths are also steadily declining from a peak of 2.2 million in 2005 to an estimated 1.8 million in 2010 while a decline of 20% was seen in the pediatric population (below 15 years) over the same time period. Regional differences contrary to these trends do however exist.

Sub Saharan Africa is home to 69% of all people (children and adults) living with HIV, with nearly 1 in every 20 adults infected in the region (HIV/AIDS Fact sheet no. 360, 2012).

The African continent, demographically a relatively young continent with 41% of the population being represented by children (410 million), has a total number of 3.1 million children who were infected with HIV by the end of 2010 (Global HIV/AIDS Response: Update 2011).

11.2 CHILDREN WITH CANCER

Cancer has recently been named the disease of the century, expecting to reach more than 20 million new cases in 2030. Most cancers are expected to occur in developing countries, the African continent being one of the most affected (World Cancer Research Fund International: Worldwide Cancer Statistics).

Extrapolating the incidence of childhood cancer from developed countries (140–150 new cases per million/year) for a population of 410 million children, an estimated 50,000–60,000 new cases of childhood cancers are expected to occur only in Africa, the continent with the highest HIV infection prevalence. The true incidence is however not known as dedicated pediatric tumor registries do not exist with the exception of the South African one.

The most common malignancies in children are represented by leukemia, brain tumors and lymphomas in developed countries (Lanzkowsky, 2005).

In developing countries the most common cancers are: non-Hodgkin lymphomas, nephroblastoma and retinoblastomas but the classification differs from region to region. In Sub Saharan Africa, HIV related malignancies are often found in the group of the three most common cancers in children.

11.3 CHILDREN WITH HIV AND CANCER

It has been shown that patients with HIV infection have a much higher chance of developing cancers directly related to their immune deficiency (Grulich et al., 2007).

While the incidence of cancer in HIV-negative children is estimated to be 0.13 to 0.15 per 1000 (Stack, 2007; Ries, et al., 1975–2004) early reports from the USA found an incidence of 2/1000 in HIV-infected children (Centers for Disease Control and Prevention, 2001; Biggar, 2000) between 10 and 20 times higher than in the general population.

Even higher figures were subsequently reported by European investigators. Before the widespread introduction of highly active anti-retroviral therapy (HAART) in Italy, the cancer rate in HIV-infected children was 4.49/1000 (Chiappini et al., 2007).

In the current era of HAART and its early initiation, cancer rates have significantly decreased in children being infected perinatally (Chiappini et al., 2007). In a Spanish study it was shown that HAART was effective at reducing AIDS defining malignancies but the rate of non-AIDS defining malignancies increased. Despite the use of HAART, the incidence of cancer appeared still to be higher in HIV-infected children compared to the general population (Alvaro-Meca, 2011).

For Africa, the higher rates of malignant disease in children with HIV are compounded by the huge numbers of those infected. In 2010, out of a total of 3,400,000 children younger than 15 years living with HIV, and 3,100,000 were resident in sub-Saharan Africa. While comprehensive prevention and treatment strategies have reduced and stabilized HIV prevalence in Europe and North America, only 21% of children in sub-Saharan Africa who need HAART are receiving treatment (Report on the Global AIDS Epidemic, 2008; UNICEF, 2010).

In South Africa, according to the mid-2011 population statistics 31.3% of the population is younger than 15 years. The estimated overall HIV prevalence rate among adults and children are approximately 10.6%. There were an estimated 63,600 new HIV infections among children aged 0–14 years (Statistical Release P0302 South Africa Mid-Year Population Estimates, 2011).

Using the published European estimates, South Africa should expect at least 1,000 new cases of HIV related malignancies in children per year (Stack, 2007; Chiappini et al., 2007)

The decrease in cancers related to HAART is significant only after 2 years of starting the treatment (Chiappini et al., 2007) It appears that although HAART has increased the life expectancy of children with HIV it has not reduced the burden of malignant disease associated with the virus.

11.4 MALIGNANCIES IN HIV POSITIVE CHILDREN: EPIDEMIOLOGY AND CLINICAL PRESENTATION

11.4.1 EPIDEMIOLOGY

Among pediatric HIV positive patients, cancers are classified as either AIDS defining or HIV associated. AIDS defining cancers are Kaposi sarcoma and Non Hodgkin lymphoma (defined as WHO IV or CDC category C conditions) while HIV associated diseases includes leyomyosarcoma (children), anal cancer, Hodgkin's disease, hepatoma and squamous conjunctival carcinoma, mainly in the adult population.

1) Kaposi Sarcoma in Children

Kaposi sarcoma was first noted in the early 1980s and became the first AIDS- defining illness. Since the development of antiretroviral treatment (ART), KS has become a rare disease in rich resource settings, especially in pediatric patients. However, KS remains the most common HIV related malignancy in Sub Saharan Africa where HIV is most prevalent and human herpes virus 8 (HHV8) is endemic (Martellotta et al., 2009). Kaposi sarcoma is a mesenchymal tumor involving blood and lymphatic vessels of multifactorial origin. There are four recognized clinical variants of the disease including classic, endemic, iatrogenic and epidemic KS. Classic KS is a rare and mild form of the disease first described in 1872 by the Hungarian dermatologist Moritz Kaposi as a vascular tumor affecting the lower extremities of elderly men of Mediterranean, east European, or Jewish heritage. Endemic or African KS is a variant of disease affecting HIV seronegative children and adults in some areas in sub-Saharan Africa. Iatrogenic KS is associated with use of immunosuppressant drugs, and in patients with autoimmune disorders, inflammatory conditions or solid organ transplantation. Epidemic or AIDS-associated KS is a more aggressive form of this disorder. It occurs mainly in the context of advanced immunosuppression even though it may develop throughout the entire spectrum of HIV disease (Hengge et al., 2002).

The current belief is that the initiation and progression of KS is a confluence of viral oncogenesis by human herpes virus-8 (HHV-8) also known as Kaposi's sarcoma associated virus (KSHV), and cytokine induced growth together with some form of immune compromise.

The HHV8 virus was discovered recently by the Chang Moore in 1994 and it was a major breakthrough discovering 2 new small fragments of DNA (novel herpes viral genome) that were present only in AIDS-KS specimens but absent in non KS tissues.

Their findings also suggested that the association of the virus with KS was not the result of HHV-8 reactivation in immunodeficient hosts because the HIV negative KS patients were not significantly immunosuppressed.

The seroprevalence of KSHV among the general population varies geographically. The precise mode of its transmission though not clearly understood is believed to include vertical, horizontal through sex or oral shedding, blood transfusion, and injection drug use, as well as solid organ or bone marrow transplantation (Fatahzadeh, 2012). In areas where KSHV infection is endemic the infection is thought to be acquired in childhood from seropositive family members and seroprevalence rates increase with age reaching as high as 80%.

In immunocompetent children, HHV8 may be associated with a febrile maculopapular skin rash while in HIV-infected older children a transient angiolymphoid hyperplasia occurs as part of the HHV-8 seroconversion syndrome. In children it is believed that KS is a manifestation of primary infection, where as in older children KS occurs after primary infection.

2) Non-Hodgkin Lymphoma in Children (NHL)

The relative risk of NHL in the HIV population is up to 1200-fold higher than in non HIV-infected children. These tumors arise because of failure of their immune system to eradicate the lymphocytes latently infected with Epstein Barr virus (EBV). This virus is known to cause malignant transformation of lymphocytes that harbor it by causing unrestricted proliferation.

Burkitt's lymphoma (BL) was first described by Dennis Burkitt in 1958 and it is known as a highly proliferative B-cell tumor.

It includes three variants: endemic (affecting children in equatorial Africa and New Guinea), sporadic (children and young adults throughout the world), and immunodeficiency related (primarily in association with HIV infection).

The key factor in the oncogenesis of BL is the activation of the c-myc oncogene on chromosome 14 of the patients who are already co infected with EBV

Diffuse large B cell lymphoma (DLBCL) is the most commonly occurring B cell NHL among HIV-infected children in whom it tends to be more indolent. Other histological forms of B-NHL include immunoblastic lymphoma and primary CNS lymphomas (PCNSL)

3) Leiomyosarcoma

Leiomyosarcomas are the second most common malignancies seen in HIV-infected children in the developed world (Mueller, 1999). The relative risk that these smooth muscle lesions will develop in an HIV-infected child compared with and HIV-infected child is 10,000. In the setting of HIV infection this tumor appears to be associated with EBV infection (McClain et al., 1995). In these children most lesions have been found in various anatomical locations including the gastrointestinal tract, liver spleen, lung and CNS. The pulmonary lesions are often visible as nodules on chest CT, whereas the GIT tumors present with evidence of obstruction, abdominal pain, and blood y diarrhoea. The course of disease is highly variable with indolent tumors (more likely leiomyomas) that probably do not necessitate intervention in some children and very aggressive, disseminated tumors in others.

Despite being described in the literature published in high-income settings this tumor is not commonly found in the African population making it difficult to clarify its association with the HIV infection (Stefan et al., 2011).

11.4.2 CHANGES IN INCIDENCE OF CHILDHOOD CANCERS ASSOCIATED WITH HIV INFECTION

The number of incidental malignancies in HIV positive children is also raised compared to HIV negative but a better overall survival in the HIV pediatric cohort has to be kept in mind.

Among African HIV positive children, the incidence of malignancies has been changing over the years. In a Zambian study a significant increase was found in Kaposi sarcoma and retinoblastoma, with a gradual sustained increase in non-Hodgkin's lymphoma, nasopharyngeal carcinoma and rhabdomyosarcoma and reduction in Burkitt's lymphoma (Chintu et al., 1995).

In a Malawian study, only Kaposi Sarcoma and non-Burkitt, non-Hodgkin lymphomas were associated with HIV infection, while the endemic Burkitt lymphoma was not significantly associated with HIV (Mutalima et al., 2010). A similar Ugandan study among children at a local pediatric HIV clinic found that 1.7% of children presented with a malignancy (Tukei et al., 2011). Kaposi sarcoma was the most prevalent malignancy occurring at an almost equal frequency between sexes whereas NHL was predominantly seen among young boys. Immune recovery after starting HAART in this series compared favorable to children without malignancy despite the use of chemotherapeutic agents. A low CD4 count did not increase the probability of death.

In a South African study published in 2011 out of 882 children diagnosed with cancer, 4% were HIV infected (Stefan et al., 2011). Kaposi sarcoma was once again the most common cancer associated with HIV infection followed by Burkitt lymphoma. The probability of developing

BL was found to be over 46 times higher in children living with HIV. This finding confirms the increased probability of developing BL, which is applicable equally to the endemic as well to the sporadic form. There were no reported cases of leiomyosarcoma in this series. KS occurred exclusively in the HIV-positive group.

No other cancer type among children has been linked with HIV infection, either in that study, or in other studies from Africa.

It is known that there is a higher incidence of cancer in males at all ages (Cook et al., 2011) but this ratio is even more augmented in the HIV infected population – due to the association with Burkitt lymphoma, a cancer affecting mainly the males

11.4.2.1 GENERAL CLINICAL PRESENTATION OF CHILDREN WITH CANCER AND HIV

The clinical presentations of the HIV related malignancies differ from region to region and are dependent on the immunity of the patient.

Skin infections are common in HIV infected children and can be classified as infections and infestations, inflammatory conditions, tumors and antiretroviral related. Children infected with HIV present with infectious skin lesions that are common in the general population but more severe. With the introduction of HAART there has been a decline in the prevalence of skin conditions associated with HIV/AIDS. Contrary to that there has been a 100-fold increase in drug reactions affecting the skin and immune reconstitution has led to a recurrence of many infectious and non-infectious skin conditions that would have been dormant (Mankahla and Mosam, 2012).

The most common site of presentation in sporadic BL is the abdomen, followed by head and neck and bone marrow. In the endemic form the jaw remains the defining clinical sign.

11.4.2.2 CLINICAL PRESENTATION KAPOSI SARCOMA

Epidemic KS has a wide spectrum of presentation ranging from minimal disease that is discovered incidentally on a routine clinic visit to aggressive growth with significant morbidity and mortality. In general the presentation may be classified into lymphadenopathic, cutaneous, mucosal, visceral and other (Hengge et al., 2002; Fatahzadeh, 2012).

Lymphadenopathic KS is the most common presentation of disease in children and it tends to occur in younger children with relatively higher CD4 counts likely owing to recent HHV8 infection with a rapid progression to malignancy (since the virus is tropic for lymph nodes during seroconversion). Lymph node involvement may be the sole presentation of disease. Massive lymph node enlargement and lymphoedema may develop.

Cutaneous lesions vary in size characteristics and number ranging from a small number of isolated lesions to widespread cutaneous involvement. The plaque-like lesions occur on the thighs, calves or soles of the feet and may be exophytic and fungating with breakdown of overlying skin. The lesions are often complicated with lymphoedema, which may occur as a relatively isolated finding and may be out of proportion to the extent of visible cutaneous disease.

Mucosal disease has been recently more and more described and it involves the oral cavity. It may also be the initial presentation of KS. Laryngeal involvement may occur, the most common site being the epiglottis. Presenting symptoms of laryngeal KS may include pain, bleeding, dysphagia speech abnormalities and airway compromise. KS involving the sclera is commonly missed.

Other organs involvement includes the lungs, gastro intestinal tract, the heart and pericardium, kidneys, urogenital tract and bone marrow. Involvement of the brain and intraorbital structures is rare likely owing to their lack of lymphatics.

11.4.2.3 CLINICAL PRESENTATION NHL

While abdominal lymphadenopathy is still the most common presentation of HIV-related B-NHLs, extra nodal disease occurs relatively more frequently among this population compared to HIV-negative children and involves sites such as the CNS, bone marrow, sinuses, adrenal gland, heart, lungs and mediastinum. Disease tends to be aggressive and the children will commonly present with advanced disease involving the central nervous system and the bone marrow.

11.5 RECENT ADVANCES IN THE STUDY OF HIV POSITIVE CHILDREN WITH CANCER

It remains a challenge to investigate the relationship between HIV and cancer in children. Complicating factors are socio-economic conditions, the use of non-standardized HAART regimens and poor adherence to either ART or chemotherapy in most of the developing countries. When looking at long-term cancer risk among people diagnosed with AIDS during childhood, a continued elevated risk exists into adulthood for two AIDS defining cancers (KS and NHL) as well as the HIV associated malignancy, leiomyosarcoma. Despite this risk once HAART has been initiated, a dramatic decline was demonstrated in the AIDS defining cancers of KS and NHL (Simard et al., 2012).

Several studies in Africa have looked at the outcome and presentation of Non-Hodgkin lymphoma in the HIV pediatric population. Studies specifically done in Uganda, Africa after 2009 have noted a significant number of children with HIV and Burkitt Lymphoma (Newton et al., 2001), whereas this association has been known for longer in the adult population (Wabinga et al., 2000). Children present with lymphadenopathy in contrast to the endemic form of BL and also at a more advanced stage of disease (many cases of abdominal tumors and liver involvement). Irrespective of immune status, treatment response to chemotherapy seemed similar; suggesting a common disease process, but HIV infection had a significantly negative impact on survival of children (Orem et al., 2009).

The South African study of 882 children found a lack of association between HIV and acute lymphoblastic leukemia (ALL) (Stefan et al., 2011). Contrary to the "hygiene hypothesis" of childhood ALL developing as a response to delayed infection after inadequate immune system exposure and modulation (Greaves, 1997), more recent evidence are challenging this, and the South African study could not demonstrate such an indirect link between HIV-related immunosuppression, increased infections and ALL. The same study concludes that while these findings do not support the hypothesis of an underlying infectious cause, neither do they completely refute it. It is possible that children with HIV infection die before the development or diagnosis of overt leukemia.

11.5.1 TREATMENT OF CANCER IN CHILDREN WITH HIV

HIV-infected children with associated immunosuppression have in many cases other secondary pathology to be kept in mind, specifically in Africa and parts of Asia: tuberculosis, worm infestations, malaria and bilharzias. In contrast it is rare for HIV negative children to be diagnosed with tuberculosis and malignancy simultaneously (Davidson, 2010). Malignancy and tuberculosis in a high incidence setting present together frequently and the diagnosis of one does not exclude the other one.

Malnutrition in children with cancer is frequently seen and not necessarily related to AIDS. During cytostatic therapy, the malnourished children have a higher rate of profound neutropaenia, resulting in a higher risk of severe infections and death (Israëls et al., 2009).

The risk of Immune Reconstitution Inflammatory Syndrome (IRIS) when initiating antiretroviral treatment at the same time as chemotherapy led to the consideration of reducing the intensity of treatment regimens in pediatric cancer. Such a reduced intensity approach should also be considered in children presenting with severe malnutrition or other forms of immunosuppression upon starting chemotherapy.

In the literature there are well-established guidelines and protocols for the treatment of childhood cancer. Cancer therapy is associated with a severe immune suppression due to the destruction of leukocytes.

In HIV positive children with cancer, chemotherapy induces a much longer lasting leucopoenia than in HIV negative subjects and thus exposes them to potentially fatal infections. HAART was seen to promote faster remission from leucopoenia, thus administering a full dose chemotherapy regimen in these patients appears justified. The extent to which HAART restores the immune response to a level comparable with HIV negative children, during cancer chemotherapy is insufficiently studied. In adults, there are several small studies indicating comparable results of chemotherapy in HIV negative and HIV positive patients on HAART (Blinder et al., 2008; Suzuki et al., 2010).

Further, the potential interactions between cancer chemotherapy drugs and HAART are not fully understood: both drug categories are metabolized by the same liver enzymatic systems (mainly cytochrome P450) and their competing actions on the liver enzymes may result in either stronger or weaker systemic effects of some of the drugs (Mounier et al., 2009).

Despite the persistent leucopenia and prolonged recovery time, standard protocols of full intensity are preferred instead of low dose protocols just with a few exceptions mentioned previously.

An intensive prophylactic antibiotic therapy should be planned with supportive care.

Radiotherapy should be given in standard doses, irrespective of the HIV status of the patient despite an increase in toxicity, mainly for the mucosa of the digestive tract.

The use of chemotherapy drugs and antiviral therapy combined with possible antituberculosis medication, antibiotics, antifungal, anti emetics, etc. requires careful observation and monitoring of the cumulative side effects.

Newer drug formulations have been developed in the last few years for the treatment of BL in HIV positive patients such as Rituximab and Paclitaxel for the treatment of KS but unfortunately their use is not yet uniform due to the lack of clinical trials and prohibitive cost.

11.6 OVERALL SURVIVAL

Survival rates of children with cancer and HIV, treated with standard protocols and HAART are comparable to those of children without HIV, however studies found the rates of disease-free survival were significantly lower in children with cancer and HIV (Stefan et al., 2011). The percentage of HIV positive children who died due to toxicity of treatment was almost 60 times greater than in the HIV negative children. HIV positive children with cancer remain more susceptible to complications and require close monitoring with immediate intervention, should such complications be detected.

The outcomes of cancer treatment in this population of children are much improved for children who have access to HAART, with an estimated 5-year overall survival for the South African cohort of 45.2% for Burkitt's lymphoma, 67.4% for Kaposi's sarcoma, and 69.6% for incidental malignancy. These results are still inferior to those for HIV-negative children, where estimated 5-year overall survival for Burkitt's lymphoma usually exceeds 80% (Davidson, 2010).

11.7 CONCLUSIONS

Although the HIV infection does not cause cancer by itself, it strongly facilitates the oncogenetic action of other viruses: HHV8 in KS, EBV in NHL and HPV (adults).

Survival rates of children with cancer and HIV, treated with standard protocols and HAART are comparable to those of children without infection.

The infection with HIV worsens the prognosis of cancers, mainly due to inefficient immune system and severe neutropenia and multiple nosocomial infections.

Children with HIV and cancer require an increased effort in monitoring their complications in order to increase their survival.

KEYWORDS

- **AIDS**
- **disease**
- **epidemic**
- **HAART**
- **human immunodeficiency virus (HIV)**
- **malignancies**

REFERENCES

Alvaro-Meca, A., Micheloud, D., Jensen, J., Díaz, A., García-Alvarez, M., Resino, S., Epidemiologic trends of cancer diagnoses among HIV-infected children in Spain from 1997 to 2008. The Pediatric Infectious Disease Journal. 2011, 30, 764–768.

Biggar, R. J., Frisch, M., Goedert, J. J., Risk of cancer in children with AIDS. AIDS-Cancer Match Registry Study Group. JAMA. 2000, 284, 205–209.

Blinder, V. S., Chadburn, A., Furman, R. R., Mathew, S., Leonard, J. P., Improving outcomes for patients with Burkitt lymphoma and HIV. AIDS Patient Care STDS. 2008, 22, 175–187.

Centers for Disease Control and Prevention. HIV/AIDS Surveillance Report, 2001, 13(2).

Chiappini, E., Galli, L., Tovo, P. A. et al. Cancer rates after year 2000 significantly decrease in children with perinatal HIV infection, a study by the Italian Register for HIV Infection in Children. J Clin Oncol. 2007, 25, 97–101.

Chintu, C., Athale, U. H., Patil, P. S., Childhood cancers in Zambia before and after the HIV epidemic. Arch Dis Child. 1995, 73, 100–4, discussion 104–5.

Cook, M. B., McGlynn, K. A., Devesa, S. S., Freedman, N. D., Anderson, W. F., Sex disparities in cancer mortality and survival. Cancer Epidemiol Biomarkers Prev. 2011, 20, 1629–1637.

Davidson, A., EB. HIV and Childhood Cancer. CME. 2010, 28, 337–342.

Fatahzadeh, M., Kaposi sarcoma, review and medical management update. Oral Surg Oral Med Oral Pathol Oral Radiol. 2012, 113, 2–16.

Global HIV/AIDS Response: Update 2011. Epidemic update and health sector progress towards Universal Access. WHO, UNAIDS, UNICEF www.who.int/hiv/pub/progress_report2011/en/ (accessed May 10, 2013).

Greaves, M. F., Aetiology of acute leukemia. Lancet. 1997, 349, 344–349.

Grulich, A. E., van Leeuwen, M. T., Falster, M. O., Vajdic, C. M. Incidence of cancers in people with HIV/AIDS compared with immunosuppressed transplant recipients: a meta-analysis. Lancet. 2007, 370, 59–67.

Hengge, U. R., Ruzicka, T., Tyring, S. K. et al. Update on Kaposi's sarcoma and other HHV8 associated diseases. Part 1, epidemiology, environmental predispositions, clinical manifestations, and therapy. Lancet Infect Dis. 2002, 2, 281–292.

HIV/AIDS Fact sheet no. 360, November 2012; www.who/int/mediacentre/factsheets/fs360/en/ (accessed May 8, 2013).

Israëls, T., Van De Wetering, M., Hesseling, P., Van Geloven, N., Caron, H., Molyneux, E. M., Malnutrition and neutropenia in children treated for Burkitt lymphoma in Malawi. Pediatr Blood Cancer. 2009, 53, 47–52.

Lanzkowsky, P. Manual of Pediatric Hematology and Oncology. Academic Press Inc.; 2005:856.

Mankahla, A., Mosam, A., Common skin conditions in children with HIV/AIDS. Am J Clin Dermatol. 2012, 13, 153–166.

Martellotta, F., Berretta, M., Vaccher, E., Schioppa, O., Zanet, E., Tirelli, U., AIDS-related Kaposi's sarcoma, state of the art and therapeutic strategies. Curr HIV Res. 2009, 7, 634–638.

McClain, K. L., Leach, C. T., Jenson, H. B. et al. Association of Epstein-Barr virus with leiomyosarcomas in children with AIDS. N Engl J Med. 1995, 332, 12–18.

Mounier, N., Katlama, C., Costagliola, D., Chichmanian, R. M., Spano, J. P., Drug interactions between antineoplastic and antiretroviral therapies, Implications and management for clinical practice. Crit Rev Oncol Hematol. 2009, 72, 10–20.

Mueller, B. U., Cancers in children infected with the human immunodeficiency virus. Oncologist. 1999, 4, 309–317.

Mutalima, N., Molyneux, E. M., Johnston, W. T. et al. Impact of infection with human immunodeficiency virus-1 (HIV) on the risk of cancer among children in Malawi - preliminary findings. Infect Agents Cancer. 2010, 5, 5.

Newton, R., Ziegler, J., Beral, V. et al. A case-control study of human immunodeficiency virus infection and cancer in adults and children residing in Kampala, Uganda. Int J Cancer. 2001, 92, 622–627.

Orem, J., Maganda, A., Mbidde, E. K., Weiderpass, E., Clinical characteristics and outcome of children with Burkitt lymphoma in Uganda according to HIV infection. Pediatr Blood Cancer. 2009, 52, 455–458.

Report on the global AIDS epidemic 2008, UNAIDS http://www.unaids.org/en/media/unaids/contentassests/dataimport/pub/globalreport/2008/jc1510_2008globalreport_en.pdf (accessed May 9, 2013).

Ries Lag, M. D., Krapcho, M., et al. (eds.). SEER Cancer Statistics Review, 1975–2004, National Cancer Institute. Bethesda, MD, http://seer.cancer.gov/csr/1975_2004/, based on November 2006 SEER data (submission, posted to the SEER web site, 2007. Last accessed 9/12/2008.

Simard, E. P., Shiels, M. S., Bhatia, K., Engels, E. A., Long-term cancer risk among people diagnosed with AIDS during childhood. Cancer Epidemiol Biomarkers Prev. 2012, 21, 148–154.

Stack, M., Walsh, P. M., Comber, H., Ryan, C. A., O'Lorcain, P., Childhood cancer in Ireland, a population-based study. Arch Dis Child. 2007, 92, 890–897.

Statistical release P0302 South Africa Mid-year population estimates 2011. www.statssa.gov.za/publications/P0302/P03022011.pdf (accessed May 11, 2013).

Stefan, D. C., Wessels, G., Poole, J. et al. Infection with human immunodeficiency virus-1 (HIV) among children with cancer in South Africa. Pediatr Blood Cancer. 2011, 56, 77–79.

Suzuki, K., Nakazato, T., Sanada, Y. et al. [Successful treatment with hyper-CVAD and highly active anti-retroviral therapy (HAART) for AIDS-related Burkitt lymphoma]. Rinsho Ketsueki. 2010, 51, 207–212.

Tukei, V. J., Kekitiinwa, A., Beasley, R. P., Prevalence and outcome of HIV-associated malignancies among children. AIDS. 2011, 25, 1789–1793.

UNICEF. Children and AIDS, Fifth Stocktaking Report, 2010, http://www.unicef.org/publications/files/Children_and_AIDS-Fifth_Stocktaking_Report_2010_EN.pdf (accessed May 5, 2013). 2010

Wabinga, H. R., Parkin, D. M., Wabwire-Mangen, F., Nambooze, S., Trends in cancer incidence in Kyadondo County, Uganda, 1960–1997. Br J Cancer. 2000, 82, 1585–1592.

World Cancer Research Fund International: Worldwide cancer statistics http://www.wcrf.org/cancer_statistics/world_cancer_statistics.php (accessed May 10, 2013).

CHAPTER 12

OPTIMISING ANTIRETROVIRAL THERAPY VIA PHARMACOKINETICS

PRINCY LOUIS PALATTY

CONTENTS

12.1 INTRODUCTION

One of the most challenging diseases of the modern era is caused by Human immunodeficiency virus (HIV) that belongs to the genus of lentiviruses (slowly replicating retrovirus) of the Retroviridae family and the disease has been given the name "Acquired Immunodeficiency Syndrome" (AIDS), which can be called the last stage of a chronic illness of the immune system due to HIV infection. The virus which is believed to have jumped from non-human primates to humans has been an economic, developmental and social welfare threat against which no cure has been found yet. Antiretroviral therapy mainly consists of drugs that are capable of reducing the disease burden on the infected individuals and helps in preventing opportunistic infections that are often the cause of death among HIV patients. The groups of drugs range from entry inhibitors to maturation inhibitors, which aim at terminating the further development of HIV by acting at different stages of its life cycle.

12.2 DISEASE BURDEN

HIV/AIDS is the sixth biggest cause of death in the world and was responsible for 2.0 million deaths in 2004, 22 lakhs being children. Africa is the most affected continent and currently about 67% of the reported cases are from Africa. HIV has such a huge impact on the lives of Africans that in 2004, life expectancy at birth was 49 yrs and without AIDS it would have been 53 yrs. Except Africa, men are more affected than women. The disease is mainly seen among female sex workers and homosexual men.

1. After Sub-Saharan Africa, South East Asia has the most number of HIV infected individuals, 80% of whom lives in India. The spread of HIV infection has been so rapid in India that it has now spread to all states. The number of affected individuals was estimated to be 2.31 million at the end of 2007. HIV prevalence among adults (15–49 yrs) is less than 1% and is greater among males (0.44%) than females (0.23%). In six states – Andhra Pradesh, Karnataka, Manipur, Maharashtra, Nagaland, and Tamil Nadu – the HIV epidemic is generalized. In India, highest prevalence is among CSWs and their clients, men having sex with men (MSM),

intravenous drug users (IDUs), truck drivers and transgendered individuals. Both HIV-1 and HIV-2 are found in the country, HIV-1 subtype C being the predominant one (Melanie et al., 2012).

12.3 HISTORY

Two types of HIV have been characterized: HIV1 and HIV2. Both are believed to have originated in West-Central Africa and are clearly recognized as viral zoonotic infections. Studies suggest that HIV1 entered the human population in the early 20th century; probably between 1915 and 1941 having evolved from a Simian Immunodeficiency Virus (SIVcpz) found in the chimpanzee subspecies Pan troglodytes. HIV2 has evolved from a different strain of SIV, found in the sooty mangabeys of Guinea-Bissau, Gabon and Cameroon. The virus could have been transmitted through various ways including bites, scratches, or medicinal preparations since monkeys in the above mentioned geographical areas are kept as house hold pets and virtually all of them were born in the wild. Monkeys are also valued food material in certain parts of Africa. The rapid cultural, political and social changes in the African continent could have played a role in the rapid spreading of HIV.

In India, the first known case of HIV was diagnosed in 1986 by Dr. Suniti Solomon amongst female sex workers in Chennai. Within a year 135 more cases were discovered among which 15 had already progressed to AIDS. Since then the transmission of HIV infection in India has been so rapid that in 2010, as per UNDP's report, 2.39 million people are infected with HIV. The Indian government set up the National AIDS Control Programme in 1987 mainly to coordinate national responses such as blood screening and health education. In 1992, NACP Phase I was implemented with objectives of reducing the morbidity and mortality rate and also to control the spread of HIV infection across the country. Phase II (1999–2006) was aimed at resisting the epidemic on a long-term basis whereas Phase III (2007–2012) mainly concentrates on preventive measures, support for the patients and treatment.

The first antiretroviral drug to be introduced was Zidovudine, also called azidothymidine (AZT) in 1984 which was initially rejected due to its toxicity and very unpleasant side effects although twenty years later it has made a comeback and is now included among the basic antiretroviral drugs. Over these years many types of antiretroviral drugs have been introduced which are capable of increasing the quality of life and lifespan of the patients but are incapable of destroying the virus and thus the disease as such.

12.4 REPLICATION OF HIV

Once the virus gains entry into the human system by any route; sexual, blood transfusion or mother to fetus transmission, it passes through various stages of its life cycle before it destroys even a single cell.

These stages are as given below:

1) Selective tropism for CD4 molecule receptor: After entering the host body, the virus develops affinity for cells having CD4 receptor molecule on their surface, the most importanT-cell being CD4+T-cell also called helper T-cell. Other cells having these receptor molecules but less affected by the virus include dentritic follicular cells, monocyte-macrophages, microglial cells, epithelial cells of cervix, Langerhans cells of skin.
2) Internalization: With the help of chemokine coreceptor (CCR), gp120 of the virion fuses with the CD4 molecule receptor of the hosT-cell (CD4+T-cell), following which gp41 glycoprotien of the envelop gets internalized into the CD4+T-cell.
3) Uncoating and viral DNA formation: The virus undergoes uncoating once it is inside the host cytoplasm. Reverse transcriptase synthesizes single stranded DNA from the viral RNA. DNA polymerase then converts it into double stranded DNA, destroying the original viral RNA. The viral DNA so produced undergoes repeated mutations, making them quite resistant to antiviral therapy.
4) Viral integration: It refers to the integration of the viral DNA into the host DNA present in the nucleus with the help of viral integrase protein. The viral particle at this stage is called HIV provirus.
5) Viral replication: The viral DNA that has been integrated into the host DNA transcripts viral RNA which is then taken into the cytoplasm where they acquire protein coat. The multiplication is further facilitated by cytokines released by the CD4+T-cell (helper T-cell).

6) Latent period and attack by immune system: In a short latent period of the infection, the host immunity tries to attack the virus in the form of CD4+T-cell, CD8+T-cell (cytotoxic T-cell), macrophages and by formation of antibodies. But the defense is soon overpowered by the virus.
7) HosT-cell destruction: The viral particles in the cytoplasm of hosT-cell form buds from the hosT-cell wall causing damage to the hosT-cell. When the damage is severe, the hosT-cell is killed by apoptosis.
8) Viral dissemination: Once the hosT-cell ruptures, the viral particles are carried through blood and lymph to infect other CD4+T-cell. The viral particles also lodges in the lymphoid tissues which form the major reservoir of the virus.
9) Impact on other cells: Monocyte-macrophages, dentritic follicular cells are a few other cells that can be infected by HIV virus. Monocytes and macrophages do not get destroyed but becomes a reservoir of the virus. The infected dentritic follicular cells causes massive enlargement of follicular centers resulting in generalized lymphadenopathy (Tsiodras et al., 2000).
10) Infection of nervous system:

The complete treatment of HIV-AIDS involves:

- Antiretrovirals;
- Treatment of opportunistic infection;
- Anticancer drugs for Kaposis square;
- Drug stomasis to increase immunity.

12.5 DRUG THERAPY OF AIDS

12.5.1 ANTIRETROVIRAL DRUGS

a) Nucleoside reverse transcriptase inhibitors (NRTIs):

Mechanism of Action:

NRTIs act by terminating the elongation of the viral DNA that has been integrated into the host DNA, thus preventing the production of new virions.

Name of the drug	Pharmacokinetics	Adverse effects	Interactions
Zidovudine	Bioavailability is 65% Plasma t1/2 of 1 hr Plasma binding is 30% Crosses placenta and found in milk.	Low therapeutic index, nausea, anorexia, abdominal pain, headache, insomnia, myalgia.	Paracetamol, Azole antifungals increase toxicity. Stavudine and zidovudine exhibit mutual antagonism.
Didanosine	Bioavailability is about 42% Plasma t1/2 is 1.5 hr	Peripheral neuropathy, pancreatitis, nausea, vomiting, abdominal pain.	Interactions seen with lamivudine, abacavir, sildenafil and tenofovir
Stavudine	Bioavailability is 99% in adults and 78% in children. Plasma t1/2 of 1.5 hr	Peripheral neuropathy, pancreatitis, nausea, vomiting, abdominal pain, bone marrow suppression.	Should not be administered with zidovudine due to antagonism.
Lamivudine	Bioavailability is Plasma t1/2 is 6–8 hrs Excreted unchanged in urine	Headache, fatigue, nausea, anorexia, pain.	May enhance the toxic effects of Emtricitabine. Ribavirin may enhance the hepatotoxic effect of Lamivudine.
Abacavir	Bioavailability is 80%, Plasma t1/2 is 1.5 hrs	Hypersensitivity reactions such as rashes, fever, flu like symptoms.	Protease inhibitors may decrease the serum concentration of Abacavir. Ribavirin may enhance the hepatotoxic effect.

b) Non-nucleoside reverse transcriptase inhibitor (NNRTIs):

Mechanism of Action:

NNRTIs inhibit the viral reverse transcriptase enzyme by directly binding to it and thus preventing the formation of double stranded viral DNA.

Name of the drug	Pharmacokinetics	Adverse effects	Interactions
Efavirenz	Bioavailability is 50% Plasma t1/2 is 48 hrs Totally metabolized	Headache, rashes, dizziness, insomnia, other neuro-psychiatric symptoms	CNS depressants like alcohol will enhance the CNS side effects. Decreases the serum concentration of Apixaban, Atazanavir. Increases the serum concentration of budesonide, Carvedilol.
Nevirapine	Well absorbed orally. Plasma t1/2 of 30 hrs Extensively metabolized in the liver.	Rashes, nausea, headache, fever, hepatotoxicity.	Decreases the concentration of oral contraceptives, Apixaban, Lopinavir, Saquinavir. Increases the serum concentration of Atazanavir, Clarithromycin.
Delavirdine	Oral bioavailability is 85% Plasma t1/2 is 6 hrs Extensively metabolized in the liver Protein binding is ~98%.	Rashes, nausea, vomiting, depressive symptoms.	Antacids decrease the serum level of Delavirdine. Delavirdine increases the serum levels of Budesonide, Fluticasone, Diclofenac, Citalopram, etc.
Etravirine	Bioavailability is 50% increased with food. Plasma t1/2 of 41 hrs. Extensively metabolized in liver. Excreted in the feces.	Rashes, nausea, increased level of LDL cholestrol	Decreases the serum concentration of Amiodarone, antifungal agents. Increases the serum levels of Atazanavir, Carvedilol, Digoxin, Citalopram.
Rilpivirine	Bioavailability is 40% increased with food. Plasma t1/2 is ~50% Protein binding: 99.7% Hepatic metabolism, excreted in feces.	Increase in levels of LDL, total bilirubin, ALT, AST, abdominal pain, depressive symptoms.	Antacids and carbamazepine may decrease the serum concentration of Rilpivine. Didanosine decreases the absorption. Rilpivine increases the metabolism of methadone.

c) Retroviral protease inhibitor (PI):

Mechanism of Action:

HIV proteases catalyze the proteolytic cleavage of the polypeptide precursors into mature enzymes and structural proteins. Protease inhibitors prevent the further development of budded immature viral particles that contain catalytically inactive proteases and thus they cannot develop into infective form.

Name of the drug	Pharmacokinetics	Adverse effects	Interactions
Indinavir	Good bioavailability Plasma t1/2 of ~2 hrs Protein binding: 60% Hepatic metabolism, excretion in feces.	GI intolerance, hyperbilirubinemia, nephrolitiasis, headache, dizziness, pruritis, rash	Decreases plasma concentration of Abacavir, Amioderone. Antacids decrease the absorption. Increases the serum concentration of Avanafil, Budesonide. Carbamazepine increases the metabolism.
Nelfinavir	Food increases the bioavailability. Plasma t1/2 of 5 hrs. Protein binding: >95% Hepatic metabolism, excretion in feces.	Diarrhea, rashes, nausea, flatulence	Antacids decrease the absorption. Nelfinavir decreases the serum concentration of Abacavir, oral contraceptives. Increases the serum concentration of Budesonide, calcium channel blockers, corticosteroids.
Ritonavir	Bioavailability increases with food intake. Plasma t1/2: 3–5 hrs Protein binding: 99% Hepatic metabolism, excreted in urine	Hypercholesterolemia, nausea, vomiting, diarrhea, headache, insomnia, rashes	Antacids decrease the absorption. Ritonavir decreases the serum concentration of Amioderone, oral contraceptives. It diminishes the therapeutic effect of Clarithromycin, codeine.
Saquinavir	Poor bioavailability Plasma protein binding: ~98% Extensively metabolized in liver, excreted in feces.	Nausea, chest pain, fatigue, anxiety, rashes, pruritis	Antacids decrease the absorption. Carbamazepine increases the metabolism. Saquinavir decreases the serum concentration of budesonide, fluticasone, diazepam.

Name of the drug	Pharmacokinetics	Adverse effects	Interactions
Lopinavir	Protein binding: 99% Plasma t1/2: 5–6 hrs Hepatic metabolism, excreted in feces.	Hypercholesterolemia, nausea, vomiting, diarrhea, headache, insomnia, rashes	Antacids decrease the absorption. Lopinavir decreases the serum concentration of Amioderone, oral contraceptives. It diminishes the therapeutic effect of Clarithromycin, codeine.
Tipranavir	Plasma protein binding: >99% Bioavailability not established. Plasma t1/2: 5–8 hrs Hepatic metabolism, excreted in feces.	Rashes, hepertriglyglyceridemia, fever, fatigue, dehydration.	Enhances the effect of antiplatelet agents. Decreases the serum concentration of Abacavir, Amioderone. Antacids decrease the absorption.
Fosamprenavir	Bioavailability: not established Protein binding: ~90% Plasma t1/2: ~7.7 hrs Rapidly converted to amprenavir which is then metabolized by liver. Excreted in feces and urine.	Rashes, hypertriglyceridemia, diarrhea, headache, fatigue	Antacids reduce the absorption. Increases the serum concentration of Budesonide, Calcium channel blockers. Decreases the serum concentration of Abacavir.
Darunavir	Bioavailability: 82% Plasma t1/2: 15 hrs Protein binding: ~95% Hepatic metabolism, excreted in feces and urine.	Hypercholesterolemia, vomiting, diarrhea, headache, fatigue, rashes.	Increases the serum concentration of Budesonide, Calcium channel blockers. Decreases the serum concentration of Abacavir.
Atazanavir	Absorption increases with food intake, Protein binding: 86% Plasma t1/2: 7–8 hrs Hepatic metabolism, excretion in feces and urine.	Rashes, increases cholesterol and bilirubin levels, abdominal pain, vomiting, diarrhea, thrombocytopenia	Antacids reduce the absorption. Increases the serum concentration of Budesonide, Calcium channel blockers. Decreases the serum concentration of Abacavir.

d) Entry inhibitors:
MECHANISM OF ACTION:
These drugs prevent the attachment of viral gp120 envelope protein either to the CD4 receptor molecule or to CCR5/CXCR4 co-receptors.

Name of the drug	Pharmacokinetics	Adverse effects	Interactions
Enfuvirtide	Bioavailability: ~84% Protein binding: 92% Metabolism by proteolytic hydrolysis Plasma t1/2: ~4 hrs	Fatigue, insomnia, diarrhea, nausea, folliculitis	Enfuvirtide increase the plasma concentration of protease inhibitors.
Maraviroc		Fever, rashes, upper respiratory tract infection, dizziness, insomnia	CYP3A4 Inducers decrease the serum concentration of Maraviroc.

12.6 ANTI-HIV REGIMENS

Certain points to be kept in mind while initiating a combination of antiretroviral drugs are:

- Therapy with three antiretroviral drugs are preferred.
- The three drugs in the regimen should belong to at least two different groups (preferably 2 NRTIs and a NNRTI or PI).
- Three NRTI regimen is employed when the patient shows drug interactions with NNRTI or PI.
- PI sparing regimens are better due to lower pill burden, less metabolic complications and to avoid higher toxicity.
- The drugs should not be discontinued at any cost, even though it will provide temporary relief from the adverse effects only to increase the viral load. Development of drug resistance is also seen in these patients.
- Treatment is life-long and pregnancy does not contraindicate antiretroviral therapy.

12.6.1 PREFERRED REGIMENS

a) 2 NRTI+NNRTI (PI sparing): Zidovudine + Lamivudine + Efavirenz
b) 2 NRTI+PI: Zidovudine + Lamivudine + Lopinavir

12.6.2 ALTERNATIVE REGIMENS

a) 2 NRTI+NNRTI (PI sparing):
   - Zidovudine + Lamivudine + Nevirapine
   - Lamivudine + Stavudine + Efavirenz
   - Lamivudine + Stavudine + Nevirapine
   - Lamivudine + Abacavir + Efavirenz
   - Lamivudine + Abacavir + Nevirapine

b) 2 NRTI+PI:
   - Lamivudine + Zidovudine + Indinavir
   - Lamivudine + Stavudine + Ritonavir
   - Lamivudine + Abacavir + Lopinavir
   - Lamivudine + Abacavir + Nelfinavir

c) 3 NRTI:
   - Zidovudine + Lamivudine + Abacavir

12.7 INDIVIDUAL ANTIRETROVIRALS

12.7.1 ZIDOVUDINE

Also known as AZT (Azidothymidine), Aztec, Novo-Azt, Retrovir is one of the widely used antiretroviral drugs, also the first one to be used as an antiretroviral drug. It is structurally related to the nucleoside thymidine. Zidovudine has an azido group at the 3' position of the deoxyribose ring whereas in thymidine it is hydroxyl group. It is a prodrug which is activated in the lymphocytes by phosphorylation. The dosage varies with age group and pregnancy. In adults oral dosage is 300 mg twice daily or 200 mg three times a day. In some patients Zidovudine has been reported to cause severe anemia and deadly hypersensitivity reaction.

12.7.1 DIDANOSINE

Didanosine is a dideoxynucleoside compound. It prevents the formation of phosphodiester linkage which is required for nucleotide chain formation. It also prevents DNA replication by binding to reverse transcriptase. It is metabolized to its active form dideoxyadenosine triphosphate intracellulary by various cellular enzymes. It is then further metabolized hepatically to yield hypoxantine, xantine and uric acid. Dosage depends on the body weight of the patient. This rule is applicable to all the age groups.

12.7.2 STAVUDINE

Stavudine is a dideoxynucleoside analog, phosphorylated into its active form stavudine triphosphate which competes for incorporation into the viral DNA. It inhibits reverse transcriptase and prevents elongation of the viral nucleotide chain. As per WHO guidelines, in adults and adolescents oral dosage is 30 mg every 12 hours regardless of body weight.

12.7.3 LAMIVUDINE

Lamivudine is a synthetic zalcitabine analong in which a sulphur atom replaces the 3' carbon atom of the pentose ring. It is phosphorylated intracellulary into its active form lamivudine triphosphate which inhibits reverse transcriptase. Lamivudine is excreted into human breast milk. Adult oral dosage is 150 mg twice a day or 300 mg once in a day. Lamivudine is also indicated in treatment of Hepatitis B infection where it inhibits HBV polymerase.

12.7.4 ABACAVIR

It is a carbocyclic synthetic analogue which is intracellulary converted into its active form carbovir triphosphate. Carbovir triphosphate is an analogue of deoxyguanosine triphospahte which is the natural substrate HIV-1 reverse transcripatase. It inhibits the reverse transcriptase by competing with the natural substrate for incorporation into the viral DNA. Adult oral dosage is 300 mg twice daily or 600 mg once a day.

12.7.5 EFAVIRENZ

It is a synthetic purine derivative originally approved for use in patients who failed therapy with zidovudine. Like zidovudine, efavirenz is also a reverse transcriptase inhibitor which is activated into its triphosphorylated form intracellularly. Compared to zidovudine, human DNA polymerase is less susceptible to the pharmacological effects of activated efavirenz. Efavirenz is metabolized initially by cytochrome P450 system into hydroxylated compounds which then subsequently undergoes glucuronidation. Adult oral dosage is 600 mg once a day. It is always used in combination with at least two other antiretroviral drugs.

12.7.6 NEVIRAPINE

It binds directly to reverse transcriptase and causes disruption in the enzyme's catalytic site. Nevirapine is active only against HIV 1 reverse transcriptase and not against HIV-2 reverse transcriptase or human DNA polymerase. It is prescribed only after the infection has become evident and along with another ARV drug such as zidovudine or didanosine. It is extensively metabolized via cytochrome P450 3A4 into hydroxylated compounds. Adult oral dosage is 200–400 mg once a day.

12.7.7 DELAVIRDINE

Like nevirapine, delavirdine also binds to reverse transcriptase directly and disrupts the enzyme's catalytic site. It is effective only against the reverse transcriptase of HIV-1. HIV-2 RT and human DNA polymerase remain unaffected. Delavirdine is metabolized into inactive compounds by cytochrome P450 3A. Adult oral dosage is 400 mg thrice a day. Always prescribed along with any other two-antiretroviral drugs.

12.7.8 ETRAVIRINE

Formerly known as TMC-125, etravirine is a diarylpyramidine analogue that binds directly to reverse transcriptase and brings about a change in the enzyme's catalytic site. It is the first second generation NNRTI to come into the market to overcome certain limitations of efavirenz and nevirapine. Unlike first generation NNRTIs, etravirine shows a higher genetic barrier to resistance. Due to its better molecular flexibility compared to first generation NNRTIs, etravirine can retain its binding affinity to the catalytic site of reverse transcriptase inspite of binding site changes induced by NNRTI resistance mutations. Adult oral dosage is 200 mg twice a day (Dong et al., 1999).

12.7.9 RILPIVIRINE

Like other NNRTIs, rilpivirine binds to the reverse transcriptase enzyme to terminate the viral DNA replication. It is effective against most of the HIV-1 patients who are resistant to other NNRTIs. This is believed to be the result of its internal conformational flexibility and the plasticity of its interaction with the binding site. Relpivirine mainly undergoes oxidative metabolism followed by sulfate conjugation. It causes cytochrome P450 3A4 enzyme induction due to which combination with other drugs should be made carefully. Rilipivirine has better tolerance and only a few CNS side effects compared to other NNRTIs. Adult oral dosage is 25 mg once a day (Lewis, 2000).

12.7.10 INDINAVIR

Indinavir inhibits HIV-1 protease enzyme. HIV-1 protease is required for cleavage of viral polyprotein precursors into functional viral proteins. This results in the formation of non-infectious, immature viral protein. Like all other protease inhibitors, indinavir is always used in combination with other two-antiretroviral drug. It undergoes hepatic metabolization to form mainly seven metabolites, one glucoronide conjugate and seven oxidative metabolites. Adult oral dosage is 800 mg every 8 hours (Sudano et al., 2006).

12.7.11 NELFINAVIR

Nelfinavir inhibits HIV-1 protease and is always used in combination with any other two-antiretroviral drugs. It has a few benefits over other protease inhibitors in a way that it is well tolerated in pregnancy, can be used in patients completely intolerant to ritonavir, has a low grade interaction with methadone and may be well tolerated in hepatitis-C virus co-infected patients. Adult oral dosage is 750 mg thrice a day (Carr et al., 2005).

12.8 COMPARATIVE PHARMACOKINETICS OF ANTIRETROVIRAL DRUGS

Antiretroviral drugs are administered as a combination of NRTIs, NNRTIs and PIs. While choosing a combination, it's always convenient to understand which drug is better than the other in terms of its absorption, bioavailability, distribution, metabolism and excretion, so as to reduce the pill burden, cost of treatment, etc. to the maximum.

Lipophilic drugs like NNRTI and PI are better absorbed. NRTIs are water soluble with the exception of zidovudine. The main drug-metabolizing enzymes are CYP3A4, 2C9, 2C19, 2D6, 1A2, 2E1, 2B6, and 2A6. Protease inhibitors cause inhibition of these enzymes, especially CYP3A4 of which ritonavir is the most potent inhibitor and saquinavir is the least.

Among PIs, Saquinavir was the first protease inhibitor to be marketed. But it has very low bioavailability and very short plasma half-life due to high systemic clearance. But when retonavir was administered with saquinavir, its bioavailability was found to be increased and plasma half-life was prolonged. The combination of saquinavir and retonavir reduces pill burden and cost of the therapy. Indinavir has a few advantages over other protease inhibitors, the most important one being its plasma protein binding of only 60%. This will enhance the antiretroviral effect of the drug. Indinavir is also better tolerated. But indinavir has limited water solubility, which may predispose to nephrolithiasis. The bioavailability and plasma t1/2 of indinavir is improved when it is administered with retonavir. This will also reduce the chances of patients developing nephrolithiasis (Thompson et al., 2010).

Delavirdine is the largest NNRTI by molecular weight. It forms a very good combination with PIs because delavirdine inhibits their metabolism and hence increasing their availability by 50–80%. Dosage of indinavir has to be reduced to 600 mg per day when it is administered with delavirdine.

Of the NNRTIs, nevirapine and efavirenz have favorable pharmacokinetics due to their long plasma t1/2. But the main disadvantage of NNRTIs is that HIV rapidly develops resistance against these drugs. Nevirapine has a smaller molecular weight and hence better bioavailability. It is widely distributed, even in the CNS. Efavirenz also has a very good bioavailability and a long plasma t1/2. Combination of indinavir and efavirenz reduces the availability of indinavir and hence its dosage has to be increased to 1000 mg every 8 hrs (Montaner et al., 2010).

Plasma half-life of NNRTIs is longer than NRTIs and PIs. Among NRTIs, lamivudine has the highest plasma t1/2 of 6–8 hrs. NRTIs require intracellular phosphorylation to be activated whereas NNRTIs do not require intracellular activation. Like NNRTIs, NRTIs also can be given once or twice a day due to the long intracellular half-lives of their active triphosphate form. With the exception of Zidovudine and Abacavir, NRTIs are excreted unchanged in the urine due to which drug interactions are rare. But, NNRTIs and PIs undergo hepatic metabolism and thus drug interactions are very common (Politch et al., 2012).

Studies suggest that better lipid profile can be achieved by switching from a lopinavir-containing regimen to an atazanavir-containing regimen while maintaining virologic suppression.

12.9 NEWER ANTIRETROVIRALS

The fusion inhibitors are of two types:

a) CCR5 co receptor antagonists;
b) Fusion inhibitors.

   The HIV on entry binds to CCR5 or CXCR4 coreceptor.

   Maruviroc is for the CCR5-tropic HIV 1 at a dose of 60 mg for 24 weeks shows a good virologic response but hepato toxicity naso-pharyngttes, fever presominate (Rieder et al., 2008).

   The fusion inhibitors enfuvirtide are now being frequently used with local injection reaction, hypersensitivity reaction and pneumonia (Islam et al., 2012).

   The integrase inhibitors Raltegravir and elvitegravir are commonly used.

   Potential targets are zinc funger motif inhibitors, HIV glycoprotein membrane domain inhibitors hosT-cell chemokine receptor inhibitors (Gupta et al., 2005).

12.10 THERAPEUTIC MONITORING OF ART

Optimal therapy with antiretrovirals requires monitoring to assess therapeutic responses and predict ADR, especially on long-term therapy. It also confirms adherence which is crucial in development of drug resistance. It is a good idea, to link pill taking with daily activity. Directly observed therapyimproves virdogic suppression.

As ART interpretation has done consequence s for opportunistic infections it is importance to note medication persistence which is defined as the duration of time from initiation to discontinuation of therapy. In adherence we took for 'how of often' while persistence indicates drug taking behavior 'how long.'

Further visit frequency and regular interesting action (total blood count, differential count) BUN, greatina, LFT) are advocated.

Virologic failure may be primary to achieve viral load <50 copies/mL or sustained recurrence of such load which may be due to resistance, interaction or failure of ART.

Viral blips is an isolated level of HIV RNA copies >50–500/mL when on a chronic stable ART regimen.

CD4 count is measure of virologic response but disconduct results are see when there is increased CD4 count despite inappropriate viral suppression in some cases decreased replication capacity decreased immune activation and preserved non syncytium including strains also the reverse is good viral suppression and poor CD4 count have been seen often with combination of didanosine and tenofovir.

The ART regimen need to be changed when virologic failure is noted. Drug toxicity and poor compliance necessitates change of therapy.

Present ART has brought down the mortality and morbidity with HIV, but the life long use of drugs is suppressed by adverse effects and adherence.

Hence the concept of structural treatment interruptions have been the goal in order to induce long-term immune control of HIV.

Moreover, STI would reduce drug toxicity, pill binders and save on cost. This is based on CD4 T-cell guided treatment interpretation of short cycles of 'on/off therapy' strategies of fixed duration.

In fact efavirenz entricilative and tenofovir is advocated for 5 days with 2 days discontinuation. The question arises in cases of virologic failure with drug resistance.

STI before initiation of optimized ART in patient with multidrug resistance does not have impaired virologic or immunologic response for treatment failure is most often due to drug resistance.

Stavudine lamivudine and envirapine is the commonly used FDC.

Cost effectiveness is another matter of paramount practical importance. Global funding is easing the problem.

Pharmacogenetics markers compredict ADR in patients is a useful tool along with TDM therapeutic drug monitoring.

12.11 CONCLUSION

- The long-term requirement of ART is a beset with problems of compliance.
- Toxicity and resistance.
- The need for simpler effective regimens and optimal combination of various ART should be effective.
- Comparative pharmacokinetics and therapeutic drug monitoring would optimize therapies.

KEYWORDS

- **antiretrovirals**
- **comparative**
- **immunologic failure**
- **optimising**
- **pharmacokinetics**
- **virologic failure**

REFERENCES

Bedimo, R., Maalouf, N. M., Zhang, S., Drechsler, H., Tebas, P. Osteoporotic fracture risk associated with cumulative exposure to tenofovir and other antiretroviral agents. *AIDS*. 2012, 26(7), 825–831.

Carr A, Samaras K, Chisholm, D. J., Cooper, D. A., Pathogenesis of HIV-1 protease inhibitor associated peripheral lipodystrophy, hyperlipidaemia and insulin resistance, Lancet 351, 1881–1883, 1998' Egger, S., Drewe, J. Interactions of Cardiac and Antiretroviral Medication. Herz 2005, 30(6), 493–503.

Dong, K. L., Bausserman, L. L., Flynn MM et al. Changes in body habitus and asrum lipid abnormalities in HIV positive women on highly active antiretroviral therapy (HAART). J Acquir Immune Defic. Syndr., 21, 107–113, 1999.

Gupta, S. K., Eustace, J. A., Winston, J. A., et al. Guidelines for the management of chronic kidney disease in HIV-infected patients, recommendations of the HIV Medicine Association of the Infectious Diseases Society of America. *Clin Infect Dis*. 2005, 40(11), 1559–1585

Islam, F. M., Wu, J., Jansson, J., Wilson, D. P., Relative risk of renal disease among people living with HIV, a systematic review and meta-analysis. *BMC Public Health*. 2012, 12(1), 234.

Krentz, H. B., Cosman, I., Lee, K., Ming, J. M., Gill, M. J., Pill burden in HIV infection, 20 years of experience. *Antivir Ther*. 2012.

Lewis, W. Cardiomyopathy in AIDS, a pathophysiological perspective, Prog Cardiovasc Disc. 2000, 43(2), 151–70.

Melanie, A. Thompson, Judith, A. Aberg, Jennifer, F. Hoy, Amalio Telenti, Constance Benson, Pedro Cahn, Joseph, J. Eron, Huldrych, F. Günthard, Scott, M. Hammer, Peter Reiss, Douglas, D. Richman, Giuliano Rizzardini, David, L. Thomas, Donna, M. Jacobsen, B.S., Paul, A. Volberding, Antiretroviral Treatment of Adult HIV Infection 2012 Recommendations of the International Antiviral Society–USA Panel. *JAMA*. 2012, 308(4), 387–402. Doi: 10.1001/jama.2012.7961.

Montaner, J. S., Lima, V. D., Barrios, R., et al. Association of highly active antiretroviral therapy coverage, population viral load, and yearly new HIV diagnoses in British Columbia, Canada, a population-based study. *Lancet*. 2010, 376(9740), 532–539.

Politch, J. A., Mayer, K. H., Welles, S. L., et al. Highly active antiretroviral therapy does not completely suppress HIV in semen of sexually active HIV-infected men who have sex with men [published online March 23, 2012]. *AIDS*.

Rieder, P., Joos, B., von Wyl, V., et al., Swiss HIV Cohort Study. HIV-1 transmission after cessation of early antiretroviral therapy among men having sex with men. *AIDS*. 2010, 24(8), 1177–1183, Hollingsworth, T. D., Anderson, R. M., Fraser, C. HIV-1 transmission, by stage of infection. *J Infect Dis*. 2008, 198(5), 687–693.

Scherzer, R., Estrella, M., Li, Y., et al. Association of tenofovir exposure with kidney disease risk in HIV infection. *AIDS*. 2012, 26(7), 867–875.

Sudano, I., Spieker, L. E., Noll, G., Corti, R., Weber, R., Luscher, T. F., Cardiovascular disease in HIV infection. Am Heart J 2006, 151 (6), 1147–55.

Thompson, M. A., Aberg, J. A., Cahn, P., et al., International AIDS Society–USA. Antiretroviral treatment of adult HIV infection, 2010 recommendations of the International AIDS Society–USA panel. *JAMA*. 2010, 304(3), 321–333.

Tsiodras, S., Mantzoros, C., Hammer, S., Samore, M. Effects of protease inhibitors on hyperglycemia, hyperlipidemia and lipodystrophy, a 5 year cohort study. Arch. Intern. Med 2000, 160, 2050–2056.

INDEX

B

C

I